# Acknowledging Indigenous Knowledge

This book explores how the landscapes in indigenous territories are rapidly changing due to increased global industrial demand. This deforestation and urbanization have isolated the indigenous peoples from practicing 'traditional ways of life.' Portrayed in this book is the indigenous people's perspective of their indigenous knowledge (IK) about the environment and why losing IK is a threat to humans, wildlife, and nature. Insight is shared into why acknowledging IK as a science can help solve climate change, food and nutrition insecurity, and increasing new types of pandemics through evidence-based stories from indigenous peoples.

Features:

- Bridges the fractured space between science and nature.
- Documents the perspectives of indigenous peoples about their ancestral knowledge.
- Provides ethnographic qualitative comparative case studies of forest-dwelling indigenous peoples over a 19-year period.
- Covers largely remote indigenous territories of ten tropical countries in the Global South.
- Provides evidence-based stories examining indigenous knowledge's role in the tropics in preserving diverse landscapes and providing nature-based solutions.

**Urbanization, Industrialization, and the Environment**
*Series Editor: Viduranga Waisundara*

*Crisis Management of Chronic Pollution: Contaminated Soil and Human Health,*
Edited by Magalie Lesueur Jannoyer, Philippe Cattan, Thierry Woignier, and Florence Clostre

*Biochar: Production, Characterization, and Applications*
Edited by Yong Sik Ok, Sophie M. Uchimiya, Scott X. Chang, Nanthi Bolan

*Phytotechnology: A Sustainable Platform for the Development of Herbal Products*
Edited by Wanderley Pereira Oliveira

*Waste Technology for Emerging Economies*
Edited by Dr. T.C. Bamunuarachchige, Professor H.K.S. de Zoysa

*Advanced Materials towards Energy Sustainability: Theory and Implementations*
Edited by Samsul Ariffin Abdul Karim and Poppy Puspitasari

*Acknowledging Indigenous Knowledge: Voices of Tropical Forest People*
By Purabi Bose

# Acknowledging Indigenous Knowledge

## Voices of Tropical Forest People

By

Purabi Bose

CRC Press
Taylor & Francis Group
Boca Raton London New York

CRC Press is an imprint of the
Taylor & Francis Group, an **informa** business

First edition published 2024
by CRC Press
2385 Executive Center Drive, Suite 320, Boca Raton, FL 33431

and by CRC Press
4 Park Square, Milton Park, Abingdon, Oxon, OX14 4RN



*CRC Press is an imprint of Taylor & Francis Group, LLC*



*Library of Congress Cataloging-in-Publication Data*
Names: Bose, Purabi, author.
Title: Acknowledging indigenous knowledge : voices of tropical forest people / by Purabi Bose.
Description: First edition. | Boca Raton, FL : CRC Press, 2024. | Series: Urbanization, industrialization, and the environment | Includes bibliographical references and index
Identifiers: LCCN 2023057694 (print) | LCCN 2023057695 (ebook) | ISBN 9780367710255 (hardback) | ISBN 9780367720254 (paperback) | ISBN 9781003153061 (ebook)
Subjects: LCSH: Ethnoscience. | Applied anthropology.
Classification: LCC GN476 .B67 2024 (print) | LCC GN476 (ebook) | DDC 304.2089–dc23/eng/20240323
LC record available at https://lccn.loc.gov/2023057694
LC ebook record available at https://lccn.loc.gov/2023057695

ISBN: 9780367710255 (hbk)
ISBN: 9780367720254 (pbk)
ISBN: 9781003153061 (ebk)

DOI: 10.1201/9781003153061

Typeset in Minion
by codeMantra

# Contents

Foreword, x

Preface, xii

Acknowledgments, xvi

Author Biography, xix

CHAPTER 1 ▪ Acknowledging Indigenous Knowledge: Introduction 1

1.1 INTRODUCTION: INDIGENOUS KNOWLEDGE 1

1.2 METHODS, CASE STUDIES, AND FRAMEWORK 3

1.3 INDIGENOUS KNOWLEDGE FROM A MULTISTAKEHOLDER CONTRARIAN PERSPECTIVE 6

1.3.1 Non-Governmental Organization 7

1.3.2 Philanthropy 8

1.3.3 Government Officials (From Ministries of Forests and Agriculture) 8

1.3.4 Artists – Indigenous Radio 8

1.3.5 Academicians (From Higher Education Institutions) 9

1.4 MOTIVATION FOR THIS BOOK 10

1.5 DECOLONIZATION ROADS TO THE WAY FORWARD 10

REFERENCES 11

CHAPTER 2 ▪ 'Nature-Based Knowledge' Aligning Science and Wisdom 13

2.1 INTRODUCTION 13
2.2 NATURE-BASED KNOWLEDGE: ANALYTICAL FRAMEWORK 15
2.3 INDIGENOUS PEOPLES LIVING IN AND AROUND FORESTS: UNITY AND DIVERSITY 18
2.3.1 Environmental Defenders 18
2.3.2 Forest Tenure 19
2.3.3 Climate Adaptation 20
REFERENCES 21

CHAPTER 3 ▪ Extractive Industries Mining Way in Indigenous and Local Communities 22

3.1 INTRODUCTION 22
3.2 ETHICS, METHODS, AND DATA COLLECTION 24
3.3 CASE STUDIES 26
3.3.1 Peru | Madre de Dios 26
3.3.2 Colombia | Popayan 29
3.3.3 India | Chhattisgarh 32
3.4 DISCUSSION AND CONCLUSIONS 35
NOTE 36
REFERENCES 36

CHAPTER 4 ▪ Food as Commodity – 'Super' Food Insecurity of Indigenous Peoples: Analysis from Asia, Africa, and Latin America 38

4.1 INTRODUCTION: FARMS TO FINGERS 38
4.1.1 Acknowledging Indigenous Food Systems Knowledge: A Conceptual Framework 40
4.2 THE STUDY AREA AND DATA COLLECTION METHOD 41

4.3 CASE STUDIES 42
4.3.1 Ladakhis, India 42
4.3.2 Orang Rimba, Indonesia 44
4.3.3 Batwas, Uganda 45
4.3.4 Quechua, Bolivia 47
4.4 DISCUSSION 48
4.4.1 Freedom to Farm 48
4.4.2 Transition Toward 'Just Commons' Food System 50
4.5 CONCLUSIONS AND RECOMMENDATIONS 51
REFERENCES 51

CHAPTER 5 ▪ Pastoralists, Nomadic Movements, and Identity in Tropical Grasslands 52

5.1 INTRODUCTION: PASTORAL NOMADIC TRIBES 52
5.2 THE STUDY AREA AND DATA COLLECTION METHODS 55
5.3 CASE STUDIES: MALDHARIS, INDIA AND BAHIMA, UGANDA 56
5.3.1 Maldharis, India 56
5.3.2 Bahima, Uganda 60
5.4 DISCUSSION 64
5.5 CONCLUSION AND RECOMMENDATIONS 66
ACKNOWLEDGMENTS 66
NOTES 66
REFERENCES 66

CHAPTER 6 ▪ Factory Schools: Erasing Children's Indigenous Knowledge and Languages 68

6.1 INTRODUCTION 68
6.2 TREND IN NON-INDIGENOUS EDUCATION SYSTEM AND LANGUAGE MAINSTREAMING 70

6.3 CASE STUDIES: LAI AND WARLI TRIBE FROM INDIA, HUAORANIS OF ECUADORIAN AMAZON, AND URO INDIGENOUS PEOPLES OF BOLIVIA 72
6.3.1 Huaoranis of Ecuadorian Amazon, Tena 74
6.3.2 Uro Indigenous Peoples of Bolivia (Andes) 76
6.3.3 Lai Tribe of Mizoram, India's Northeast Region 77
6.3.4 Warli Tribe of Maharashtra, India's Western Region 79
6.4 DISCUSSION 80
6.5 CONCLUSIONS AND RECOMMENDATIONS 82
REFERENCES 83

CHAPTER 7 ▪ Indigenous Peoples and Wildlife Coexistence in Tropical Mountains: Socio-Cultural Impact 84

7.1 INTRODUCTION 84
7.1.1 Definitions 86
7.2 THE STUDY AREA AND RESEARCH METHODOLOGY 87
7.3 CASE STUDIES: THE NATURE OF HUMAN-WILDLIFE COEXISTENCE 88
7.3.1 Ladakh's Changtang Tibetan Plateau, India 88
7.3.2 Sierra Nevada de Santa Marta, Colombia 90
7.3.3 Annapurna Conservation Reserve, Nepal 92
7.4 DISCUSSION: LISTENING TO INDIGENOUS VOICES INDIGENOUS PEOPLES AND THEIR WILDLIFE COEXISTENCE 94
7.5 CONCLUSIONS WITH RESEARCH AND POLICY RECOMMENDATIONS 95
7.5.1 Recommendations for Key Multi-stakeholder Groups 95
7.5.2 Indigenous Peoples and Local Communities 96
7.5.3 Media and Communication Organization 96
7.5.4 Civil Societies 97
7.5.5 Government Authorities 97
REFERENCES 97

CHAPTER 8 ▪ Communicating the Art of Bridging Indigenous Knowledge with Science and Policy 98

8.1 INTRODUCTION 98

8.1.1 What is Communication Science? 99

8.2 COMMUNICATING INDIGENOUS KNOWLEDGE: EUROCENTRISM AND DECOLONIZATION 101

8.3 HOW SCIENCE COMMUNICATION INFLUENCING POLICY? 103

8.3.1 Methodology 103

8.3.2 Asia: India and Nepal 103

8.3.3 East Africa: Kenya and Uganda 104

8.3.4 Latin America: Colombia, Peru, and Mexico 105

8.4 ART OF BRIDGING WESTERN SCIENCE AND INDIGENOUS KNOWLEDGE 105

REFERENCES 106

INDEX, 107

# Foreword

**Dr. Purabi Bose** has a talent for approaching issues from a fresh angle. Indigenous peoples' lives and values have received increased attention from policymakers and researchers in social sciences and forestry for a handful of decades. This interest is reflected in the allocation of research grants by funding agencies. At present, in these fields, as in many other scientific fields, the majority of researchers are male, white, and from a Western background – such as myself. The scientific establishment is looking after its own. Purabi Bose is no stranger to the obstacles placed in the way of female researchers of color from an indigenous background like hers from the Global South. Undeterred, she leads by example toward a better balance. It is an uphill struggle, but one she undertakes with tenacious determination. One of the tangible results you are holding in your hands at this very moment.

This book, *Acknowledging Indigenous Knowledge: Voices of Tropical Forest People*, is not the first to address indigenous knowledge. What sets it apart from other publications is its emphasis on the knowledge holders. In this book, like in her films, Purabi Bose provides a podium to the people concerned. By doing so, she makes the minds that contain indigenous knowledge protagonists in the narrative rather than objects of study.

Being trained as a Western researcher, Purabi Bose has a foot in both worlds. This enables her to bridge those two worlds of knowing and thinking: a world of knowledge based on experience shared among generations and a world of knowledge based on the collection of data processed through theory, a world of wisdom, and a world of science. Each has its own value, and the sum is bigger than the parts. This is the gift this book provides us, expressed in a form that we can understand.

This book bears evidence of the geographical extent of the experience Purabi Bose has gathered on the life and livelihoods of indigenous peoples. Her interest in people's livelihoods and the way these are protected or undercut by law provided the intersection of our fields of interest, which led to an inspiring exchange of views and knowledge by which I feel enriched. The book combines a diversity of evidence-based studies documenting indigenous knowledge with insights into deforestation issues, biodiversity, land-use change, pastoralism, and indigenous education from remote landscapes of Asia, Latin America, and East Africa. Though this book focuses on tropical countries in the Global South, the policy analysis, methods, and recommendations are applicable to all countries across the Earth, including the Global North.

I do not doubt that this book will be enjoyed by students and scholars in the West and the South in the social sciences, environment, and forestry. Importantly, this book will also prove relevant to readers who do not approach environmental and social issues from a professional perspective but out of an interest in the knowledge, life, and values of indigenous peoples in many parts of this beautiful but endangered planet.

**Prof. Dr. Bernd van der Meulen**
*European Institute for Food Law*

# Preface

PEOPLE CAN OBSERVE, THINK, learn, understand, remember, and share what they have seen, understood, and learned. A vast part of what we know, we have learned from others is based on the observations and experiences of yet others. The way experience is organized into knowledge and is transferred to others greatly differs between cultures separated by time or distance. In the Western world, we have come to trust the knowledge created by science through experiments, measurements, and shared through peer-reviewed publications. At the same time, we have learned to distrust knowledge that has been acquired in ways that we cannot replicate in a controlled environment and that is shared in ways not accessible to search engines. When new medicines are based on traditional remedies, we applaud the prowess of the pharmaceutical industry. If 'biopiracy' can gain an advantage using traditional knowledge, then maybe there is something in it for us as well.

This book, *Acknowledging Indigenous Knowledge: Voices of Tropical Forest People*, is my attempt to amplify the issue as such as well as to present my point of view which is shaped by being a woman scholar with indigenous ancestry and a citizen of a low-middle-income country.

For various reasons, Western scholars, funding agencies, and academia continue to dominate and represent the names of indigenous peoples. There is no doubt that indigenous knowledge can provide crucial solutions for all sorts of contemporary problems, and it could be even better when the voices of the indigenous scholars and communities themselves are heard. For example, in a referendum, a majority of voters in Australia rejected a proposal to recognize Aboriginal people in the constitution and establish a body to advise parliament on indigenous issues (as of October 2023). On the one hand, the fact that the vote took place at all could be hailed as an important landmark that occurred 235 years after the British settlement, and 61 years after Aboriginal Australians were granted the right to vote, but on the other hand, the question remains why would voting by the country's population be needed for something that should be basic indigenous rights. Many countries are yet to recognize 'indigenous peoples' as people. The United Nations Declaration on the Rights of Indigenous Peoples (UNDRIP), adopted by the UN General Assembly in 2007, recognizes that indigenous peoples have the rights to maintain and strengthen their distinct political, legal, economic, social, and cultural institutions.

Forest and land resources are not just territory; they are a warp and weft of culture, heritage, languages, food, and social relations. Many African, Asian, and Latin American countries hold customary laws and traditional leaders, but as economies grow and customary systems evolve, the existing and new institutions find themselves in uncertain forest and land-use governance arrangements. Forest policy in some of these countries is advancing toward the inclusive rights of indigenous peoples and other local communities. However, in practice, deforestation due to extractive mining, fortress conservation, and industrial plantations has left many communities displaced.

A quick fixed solution through the 'restoration' of land and forest only offers prescriptive solutions to restore the land, which comes from racism, fear, prejudices, and misconceptions about indigenous scientific knowledge and practices. The historical colonization of indigenous peoples and their way of living has suppressed traditional knowledge systems and their 'way of living.' This is despite numerous decolonization efforts, many of these agendas are again from Western scholars that are symbolic efforts rather than trying to address the root cause of power inequality and politics. Acknowledging the effects of past and current colonialism, accepting failures, and being flexible to learn from traditional knowledge are important for forest and land governance. The land and forest restoration are incomplete if it continues to impose inequities and overlooks the environmental justice and forest stewards.

The findings presented in this book are the fruit of my scientific journey through diverse landscapes and remote indigenous territories across Asia, Africa, and Latin America. While at first glance there mainly appear differences between the people and cultures I met, at closer inspection, there appear recurring struggles they all have to face.

I open this book with an introductory chapter to outline the topic as well as to provide key arguments on why indigenous knowledge matters, who are the knowledge holders, and how the debate of scientific vs. indigenous knowledge diverted attention from the main issue of recognition of survival knowledge. What is indigenous knowledge? Who holds the knowledge? How is it used? Why does the debate of science vs indigenous knowledge exist? Recently, four children who survived 40 days in the Colombian Amazon jungle after a plane crash incident explain the significance of indigenous knowledge that even the best science practices and Western adventurers and scholars cannot teach. The book focuses on themes ranging from mining to forest plantations, human–wildlife coexistence, factory schooling, pastoralism, and the food security of those people who are directly dependent on forest resources.

This book connects diverse challenges and outcomes of forest and land-use change policies. As such, it would be beneficial as a qualitative case study for students and teachers in academic universities, research institutions, government, and non-governmental organizations. Even though I have placed my emphasis and experience on case studies from tropical landscapes, almost all of the issues I deal with are generic and applicable to other landscapes and indigenous peoples. They relate as much to the mining issues faced by reindeer herding Sami people in the Arctic, food foraging Aboriginal Australians, or resource-related displacements of Native Americans. This resonates with the United Nations' Sustainable Development Goals (SDGs) to make the world a better place by 2030. This book highlights the significance of the 17 SDGs for 370 million indigenous peoples and local communities that are guardians of the forests and natural resources but have been facing marginalization, exploitation, and exclusion. The UN Agenda aims to leave no one behind. Thus, the SDGs framework provides a historic opportunity to address the challenges and ensure that indigenous men, women, and youths too are not left behind.

I have written this book with a diverse range of audiences in mind, including students, policymakers, politicians, practitioners, philanthropists, and scientists from multi-disciplinary fields to understand it. The book is to be read as a true narrative of the voices that are often secondary or ignored in the decision-making process. Written to simplify scientific communication, this book should guide you to understand basic cases of all topics involved, and if you are new to the field, to generate curiosity to explore more. I wish that this book becomes a source of inspiration for all those interested and wanting to work with indigenous peoples across the globe without borders toward creating environmental justice and safeguarding human rights.

This book is not an exhaustive case study, but a few selected cases represent some of my research work. I could have added a dozen more chapters based on my ethnographic fieldwork. However, the idea here was to showcase the indigenous knowledge matters, and the chapters of this manuscript do the justice. There may be gaps here and there, and that is to acknowledge that learning is a continuous process. I hope that many indigenous authors like me will get motivated to continue this chain of thoughts and belief that indigenous knowledge needs acknowledgment. We need more indigenous women of color from the Global South to contribute as scholars' stories about indigenous peoples who are custodians of the forest in and around tropical forests across the world. The time is right now!

**Purabi Bose**

# Acknowledgments

I AM GRATEFUL FOR MY mother's indigenous ancestry. In 1942, my mother, Latika, was born in a tropical forested landscape of West Bengal, eastern India, under British Colonial rule. Her father, a forest dweller Adivasi belonging to the Santhal tribe, succumbed when the community was displaced due to new forest protection regulations. She was just three years old, and her mother returned to her maternal home instead of joining her in-laws at the coal mines of Hazaribaug, central India. The hardship at a young age made my mother determined to distance herself from her indigenous identity by marrying a non-indigenous husband because she believed being indigenous, one gets discriminated against. Yet she could never take away her indigenous values, which she inculcated in us through stories, food, culture, art, and traditional knowledge about nature. I admire her courage and determination and thank her for giving me and my brother, Prabir Bose, a chance to explore the world. This book is dedicated to all indigenous peoples fighting for their indigenous identity.

I am honored to have colleagues who stand against discrimination in the workplace collectively. In 2023, I faced ethnic discrimination in the workplace at the Swedish University of Agricultural Sciences (SLU), campus Alnarp in Sweden. It was the first time in my two decades of international professional career in forest and people science that I faced racism in the workplace. My three witnesses from SLU and two external supporters from international institutions amplified my feeble voice. The internal investigation team concluded that it was just a 'mistake' by the department, but the team did not recognize it as ethnic discrimination. To prove that the department was inclusive and to showcase the diversity, the investigation report highlighted that the department had staff recruited from 19 nationalities. The numbers do not tell the total facts. The fact is that all of these staff from 19 nationalities are from Europe and North America (the Global North), all are white, and all celebrate a single religious festival,

Christmas! The only permanent staff who is a citizen of the Global South, with indigenous ancestry, and a woman of color, unlike the rest not grown up celebrating Christmas, is me. Diversity is not about pushing one toward an inequitable path of 'integration.' I wish people understood that denial of racism is racism itself and is even worse than acknowledging it. I, therefore, resigned.

I remain humble from this experience and thank all those who stood by me and strengthened my voice despite the Swedish 'culture of silence' where many avoid expressing their solidarity openly. This ethnic discrimination experience certainly made me more empathetic and compassionate, and to practice the art of 'forgiving.' I became aware of how the academia of the Global North is struggling just like the Global South in acknowledging discrimination of gender and ethnic diversity. Unfortunately, in this context of discrimination, the world is equal.

Thank you, Bernd van der Meulen, Björn Hånell, Alexandra D'Urso, David Kaimowitz, Katarina Eckerberg, Sara Gräslund, Ravi Prabhu, Camilla Sandström, and the International Union of Forest Research Organization (IUFRO) team particularly Cecil Konijnendijk, Alexander Buck, and John Parrotta, for being wonderful colleagues. The list is long enough to include all research partner institutions and colleagues: you know who you are! I am honored and indebted to the Royal Swedish Academy of Agriculture and Forestry (KSLA) for appointing me as an International Fellow.

This book was made possible because of numerous indigenous communities' support and warmth received from them in various countries. I became part of their family, crossing the indigenous language barriers and gaining knowledge about forests, food, wildlife, art, and bonding with nature. I am honored to experience traveling and having you all in this beautiful life. The insights, comments, and positive criticism have improved the final manuscript. Thanks also to the editorial team, including Hilary Lafoe and Sukirti Singh, and the publisher for your patience. I acknowledge that the indigenous knowledge shared in this book is those of the people who shared it with me; I am just their voice. All views expressed in this book are entirely my own.

I appreciate all the knowledge I gained and shared that has become part of this book due to various institutions with which I have been affiliated professionally since 1997. To name a few key ones to acknowledge the support are CIFOR-ICRAF (Indonesia), CIAT (Colombia), particularly CGIAR's research program Forests, Trees and Agroforestry, and NMSWDF (India), which allowed me to travel across continents – Africa, Asia, and Latin America – in the remotest forested landscapes of indigenous territories. This book is based on my two decades of experience as a social scientist in the field, working on human and social dimensions of forest and forestry and learning from diverse indigenous communities. No funding was received for writing this book. My work has also been motivated by the lack of indigenous authors, particularly women of color from developing countries, narrating the story using ethnographic and anthropological lenses from an indigenous perspective.

Felix (Felixé Bose), thank you for being a wonderful four-legged companion. We moved from being a stray hunter feral indoor-outdoor cat to becoming an indoor hunter cat during the COVID lockdown. We traveled in India from Mumbai and Goa to live abroad in Sweden in Lomma through all types of weather – cold, dark, hot, and sunny. Gerbil, aka Mickey Mousey, instead of being hunted, became your pet companion. Watching the two of you together is like a live Tom & Jerry show! Wherever we live, people know me because of you, the world's most handsome tabby cat.

To all those who trust me, love me unconditionally, share constructive feedback to make me a better human, and always find a reason to remain connected with me through thick and thin, I owe you my sincere gratitude. Thank you.

# Author Biography

**Purabi Bose,** a citizen of India with Santhal tribe/Adivasi ancestry, is a social anthropologist and environmental scientist with a research focus on forest tenure and land rights of indigenous peoples, climate adaptation, and food security in Asia, East Africa, and Latin America. Currently, she is an Associate Professor (Docent) and Senior Lecturer of Forest Policy at the Swedish University of Agricultural Sciences (SLU), Sweden.

Her professional affiliations have spanned NGOs, philanthropic, international research, academic, and think tanks with a focus on rural and tribal development, natural resource governance, and social inclusion. She is passionate about simplifying science for better communications and helps in capacity building, teaching, partnership, policy analysis, research, and publications.

Purabi holds a Ph.D. from Wageningen University and has held numerous fellowships, including those at the University of California, Berkeley, the Swedish University of Agricultural Sciences, the University of Copenhagen, and the Smithsonian Institute. She presented her research work at over 100 international conferences. She has published her research in peer-reviewed journals and in popular science media.

As a filmmaker, she is a producer and director of independent films on forests, mountains, land, and the way of living of indigenous peoples. This is done under her self-funded initiative, Landing Together. Further details about films and protagonists at www.landingtogether.weebly.com

Her pro-bono responsibilities with the International Union of Forest Research Organization (IUFRO) include as a Voting Board Member, Coordinator of Division 6 on Social Aspects of Forests and Forestry, and a member of the Congress Scientific Committee for IUFRO Congress Stockholm, 2024. Earlier she was a Council Member of the International Association of Study of Commons (IASC). She is a member of IUCN's two commissions: CEESP and CEC. In 2023, Purabi was appointed as an International Fellow by the Royal Swedish Academy of Agriculture and Forestry (KSLA).

Growing up in a below-poverty family on the outskirts of Mumbai helps Purabi to be humble. She connects with people in 10+ languages and has been an expatriate who lived in 11 countries on four continents for professional work. Often you will find her mountaineering, cooking, playing drums, doing hand-embroidery, calligraphy, and cheering as a football fan. She works hard so that Felix, a tabby tom cat that adopted her, can have a better life, and they travel the world together.

CHAPTER 1

# Acknowledging Indigenous Knowledge

## *Introduction*

### 1.1 INTRODUCTION: INDIGENOUS KNOWLEDGE

Indigenous knowledge is critical for ensuring a healthy earth. One of the biggest challenges is to define indigenous knowledge because it is like air with no defined shape and no defined volume. The knowledge that has traversed across the globe through oral traditions for time immemorial. These oral traditions have been kept alive by many indigenous communities living in diverse landscapes from drylands to mountains, rivers to forests, and from urban to remote, isolated lands. One of the important parts of keeping knowledge alive is creating stories, songs, dances, poems, music, paintings, and other creative communication formats that talk about connections to the land, history, ecological knowledge, language, and culture stories. Many scholars, including those from Western countries and those who are non-indigenous, have written about what defines indigenous knowledge. For some, it is a network of knowledge, belief, and traditions; for others, it is collective, adaptive, and culturally and/or spatially context-specific, or it is a process through which a relationship is built with the natural environment (Semali and Kincheloe, 1999; Mistry, 2009; Gadgil et al., 2021).

Therefore, there exists no single definition of indigenous knowledge. Many terms are used based on what indigenous people know and practice.

DOI: 10.1201/9781003153061-1

Western scholars have floated terms like traditional knowledge, traditional and local knowledge, or traditional ecological knowledge. Each of these terms is debated by academicians about the use of words such as 'traditional' – the way knowledge is transmitted in a community, or 'Indigenous,' indicating the autochthonous nature of knowledge, or 'ecological,' highlighting the biological diversity that forms all living beings. Living beings, as per indigenous knowledge include rivers, seas, sky, and soil, which differs from the way science would perceive them. Culture is another term synonymously understood with indigenous knowledge. It is understood that indigenous peoples are guardians of our ecosystem due to the unique languages, beliefs, and special relations they share with all living beings.

Others consider indigenous knowledge as primitive, dangerous, pagan, superstitious, ignorant, and unimportant. These narratives have even led to campaigning for some of the indigenous knowledge as non-knowledge. One of the reasons for negativity around indigenous knowledge is some have misused the term for economic and other gains by promoting practices that are not indigenous knowledge. The lack of protection for indigenous knowledge within international law makes it vulnerable to misuse, abuse, and disownment. The modern development thinkers trying to replicate the 'success stories' at a larger geographical scale criticize indigenous knowledge because of its local context-specific influence and relevance to certain social groups, which limits its scalability. All this adds to another threat to indigenous knowledge, and that is its preservation and continuation.

Although indigenous knowledge is widely recognized particularly for the protection of biodiversity (see Gadgil et al., 1993) in international instruments such as Article 8(j) of the Convention of Biological Diversity taken at the Rio Conference on Environment and Development (1992) has highlighted this process with signatories to "respect, preserve, and maintain knowledge, innovations and practices of Indigenous and local communities embodying traditional lifestyles relevant for the conservation and sustainable use of biological diversity" (Nakashima and Roue, 2002). In 2007, the United Nations Declaration on the Rights of Indigenous Peoples further attested these instruments. However, despite international commitments in practice, many Indigenous communities face the threat of either being evicted or 'integrated' into mainstream communities, denied basic food, health, and education, infringing upon human rights. For example, as of October 2023, the Indigenous Ogiek community of Maasai Mau Forest in Sasimwani and Nkareta face eviction from their ancestral lands without regard for their well-being and future. Ogiek community is not alone but one among many indigenous peoples facing the threat of losing their traditional land rights and risking losing their knowledge and identity. Many indigenous peoples (though they may not be recognized as indigenous peoples by their country) from Asia, Africa, and Latin America including Native groups from North America and Indigenous pastoralists like Saami people from Europe are affected by degradation and loss of natural resources influencing their relationship and culture and way of living.

The book will use the term indigenous knowledge without any limitation to its academic definition but rather refer to it the way indigenous communities, the traditional knowledge holders, would like to address it, in their perspective and within the context of their 'way of living'. Indigenous knowledge for rural and indigenous peoples informs their collective decision-making about basic aspects of daily life and their ecosystems including other living beings. The 'living beings' in many indigenous territories include trees, mountains, rivers, and air ecosystems, and their entities by giving them rights not as objects but as living beings. For example, the local Māori tribe of Whanganui has been demanding legal recognition of their river in New Zealand as an ancestor for over 100 years. In the world-first, this river, as their kin, was awarded a legal status. This identity will protect the river from abuse and harm, and acknowledge the perspective of indigenous knowledge to treat it as indivisible and one entity rather than an outsider's perspective of managing it based on its different resources. For Western science, this is a major shift given that its fundamental belief is that humans are at the apex of the natural world. Not just in Māori culture, but in many diverse cultures of indigenous peoples, as mentioned in this book, the concept of nature as equal to humans and as ancestors is rooted in its language, culture, traditional wisdom, and knowledge.

## 1.2 METHODS, CASE STUDIES, AND FRAMEWORK

Primary ethnographic data collected by me that provides empirical evidence for this book. I have collected primary data using a combination of social science (anthropology and sociology) qualitative methods. The methods have included indigenous knowledge processes in documenting the case studies. A conscious decision to publish the findings as collective knowledge in this book as scientific literature does not imply that for indigenous knowledge to be acknowledged it should be published in the scientific literature. The book aims to reach out to a diverse range of global readerships that have easy access to this form of publication.

The method ensured the ethical values at all stages of research from data collection, analysis, storage, and management of sensitive and personal data. In interviews and case study group discussions, substantial efforts were made to protect the participants' physical, social, cultural, and economic well-being. Therefore, the collected data shared in this book are designed to hinder the identification of any participant. Free, Prior and Informed Consent (FPIC) is a specific right granted to indigenous peoples recognized in the United Nations Declaration on the Rights of Indigenous Peoples (UNDRIP). FPIC allows indigenous peoples to engage in sharing, withholding, or withdrawing consent, at any point, regarding projects impacting their territories. The ethical dimensions of sensitive data collection have been taken into account. The research chapters rely on participants' knowledge and wisdom and this book respects their rights, interests, sensitivities, and privacy.

I have audio and visual media as a mode of documentation during the fieldwork, with prior permission from the individual(s) and communities giving consent to publish. In this book, the cover page image is the only photo used in this book from the my audio-visual collection and the community has given me prior permission to publish it. The cover photo along with me is of a group of nomadic pastoralist Changpa indigenous people from the Tibetan plateau, which is known as the Changtang region of Ladakh, India. This is one of my many self-funded fieldworks that was also one the coldest at -35 degrees Celcius and the highest altitude at about 5500 meters ethnographic data collection.

The unique comparative case-study research approach of this book helps to engage the discussion from diverse dimensions from micro to macro ranging across individual, household, community, and regional levels. Comparative case studies of indigenous peoples living in different countries across continents highlight commonalities despite the geo-political boundary contrast. The constant unit is the indigenous people living in and around tropical countries and the specific units of analysis often remained the same such as mining, education, pastoralism, or conservation. In recent years, contemporary anthropological research has engaged in multi-sited ethnography, which 'does not contrast places assumed to be unrelated; instead, it looks at linkages across place, space, and time' (as quoted in Bartlett and Vavrus, 2017, p 7). Bartlett and Vavrus (2017) explain that comparative case study approach promotes a model of multi-sited fieldwork that studies through and across sites and scales. In this book, the use of a comparative case study approach strengthens the case study research method in social science, forestry science, and also the work with indigenous peoples to showcase how similar policies or 'development' projects unfold in diverse landscapes that are connected across scale and time.

The overall framework used in this book is the rights-based approach. A human rights-based approach recognizes and protects the forest and land tenure of indigenous peoples and local communities. From the perspective of indigenous peoples, the respect for Mother Earth and all forms of life holds a deep respect. Indigenous women have played a key role in safeguarding the health of the planet and all living beings, now and for the future. Likewise, indigenous languages continue to face the risk of extinction which will impact the way indigenous knowledge will be negatively affected. One of the underlying aspects of the rights-based approach is that individual and collective rights of indigenous peoples are acknowledged as 'right-holders' rather than as 'beneficiaries'. This shifts the power of perception towards the indigenous peoples who often are categorized as 'weak', 'poor', and 'marginal' groups to being equal power holders negotiating and claiming their rights to livelihoods and forested landscaped that they have protecting and managing for generations. One of the outcomes of the rights-based approach is its inclusion in global frameworks and guidelines such as the 2030 Agenda for Sustainable Development and the Post-2020 Biodiversity Framework.

In recent decades, though there have been many agreements on the rights-based approach found in regional and international rights frameworks, it not being legally binding makes it challenging to ensure successful implementation. This book, therefore, highlights why the rights-based framework needs to be understood from indigenous peoples' perspective. Acknowledging the rights of indigenous peoples as the right-holders by the state and non-state actors including private companies, politicians, non-governmental organizations (NGOs), donors, and intergovernmental organizations share an equal moral responsibility to protect and promote advancements of rights. The rights-based approach identifies the environmental defenders as a critical actor in this framework who strive to protect the environment peacefully with the support of indigenous communities. One of the fundamental limitations of the rights-based approach is that the way it is understood continues to be a Western scientific view, often failing to understand the way indigenous people build relationships between humanity and nature – other species including rivers, mountains, or lakes. This challenge also extends to the way how international frameworks identify the collective rights of indigenous peoples could be narrow compared to the on-the-ground realities of being inclusive to all who have been traditionally accessing the forest land and its resources, for example, nomadic pastoralists. In this book, the framework used is from the perspective of indigenous peoples that recognizing rights is good, but not good enough. Rather, there is a need to guarantee access to forests and resources, and rights to govern and manage the forest by the indigenous peoples with direct financial support to them to overcome climate and top-down development challenges.

## 1.3 INDIGENOUS KNOWLEDGE FROM A MULTISTAKEHOLDER CONTRARIAN PERSPECTIVE

In this section, how indigenous knowledge is understood from a multistakeholder perspective is described. Contrarian thinking is critical to better understanding the approaches and gaps of 'Western science' or non-indigenous science and ways in which indigenous knowledge can complement (Bose, 2019). Long before scientific research findings became formalized and norms of our formal Western education systems, humans were generating, storing, and disseminating information and knowledge about the existing natural world. Over millennia, indigenous peoples around the world have maintained knowledge systems that have been tested and evolved with new direct localized human-nature experiences embedded in cultural values, languages, food systems, and spiritual knowledge. Yet, it remains isolated from Western science and often neglected by scientific researchers for its contrarian perspectives due to their lack of awareness and cultural contexts of indigenous knowledge. This book does not negate or undermine the significance of Western science, rather the objective is to highlight the urgent need to acknowledge indigenous knowledge along with other forms of knowledge to achieve holistic, equitable, and integrated sustainable development.

Based on various face-to-face interviews between 1998 and 2018 with diverse stakeholders, I capture the essence of indigenous knowledge from actors ranging from activists to artists from across the world. The below quotes from these actors are not representative in any way of their institutions and/or their representative of the sector they work for. Nevertheless, it gives an essence of what different stakeholders' contrarian perspectives are, which is an essential element for a meaningful dialogue as a way forward towards attaining the Agenda 2030.

### 1.3.1 Non-Governmental Organization

First, one of the interviews was with the Directors of a non-governmental organization in contiguous tribal districts of Western India. As social workers, they dedicated all their lives to working with Bhil Adivasis and supporting water and forest development. According to them (personal communication: Dahod in 2010), "Indigenous knowledge can be only understood by those who speak their language, live with them for decades, and respect nature like them. Foreign researchers (non-indigenous scholars) wanting to study indigenous knowledge will fail to grasp the holistic view. Even after living and working for over five decades, we would be incorrect to define indigenous knowledge in four sentences. However, for your benefit, let us define by viewing the world with love from their language, culture, food, lifestyle, gender balance, self-respect, self-sufficiency, and respect for all forms of nature. In the last few decades, we have seen that with 'development' some of these forms of knowledge are diluting due to Bollywood influence that changes their notion of the value of their knowledge. Indigenous knowledge is evolving; what could have been defined by Bhil Adivasi elders in 1950 is different from how it is defined by their grandchildren in 2010 who returned from Dubai as wage labourers with money to buy a car and to buy GM (Genetically Modified) seeds for their farm to earn more money. They still practice indigenous knowledge but an adapted version."

In Kenya, one of the activists (personal communication: Nairobi, 2014) working for an NGO on the issue of indigenous peoples' rights to traditional land explained their perspective of indigenous knowledge by stating that "For me (as an Indigenous woman) and the pastoralist communities in Masaai Mara reserve I work with every time we breathe in and out is about indigenous knowledge. How can one document every breath just like that it is impossible to tap this knowledge in written words moreover in a foreign language. Moreover, there is some knowledge that is not even in words but in sounds, in action, which one can record using technology, but what is the point of recording when the meaning can only be experienced by feeling the soil, insects, air, in drumbeats, in sounds of wildlife or thunder [...] indigenous wisdom is transferred in some forms of communications that modern science has never taught us to learn. When you do not learn those indigenous forms of communication you cannot comprehend and thus a big part of understanding indigenous knowledge is ignored in English and French or Spanish colonial languages."

### 1.3.2 Philanthropy

At the World Bank Conference on Land and Poverty (personal communication: Washington DC 2017), I interviewed two staff working for a philanthropy with a focus on indigenous peoples in the Global South. According to them, "In modern science, everything should fit the boxes provided it is tested and proved of relevance. How can Western science approve indigenous people's claim that the river is alive? This would have seemed impossible a few years ago, but this year (2017) would be a mark in our history for the first time many of our iconic lakes and rivers in the world – New Zealand, India, Bangladesh, and Colombia – are finally getting recognized in law as legal persons and/or as living entities including a range of legal rights and protections. The problem is that indigenous knowledge is not fully understood – to deny the rivers the right to flow or deny a pollution-free river is not how indigenous peoples respect it. Therefore, any half-hearted adaptation of indigenous knowledge is more dangerous and meaningless."

### 1.3.3 Government Officials (From Ministries of Forests and Agriculture)

Various Government officials were interviewed during the period 2016 and 2019 to learn about their perception of indigenous knowledge. One of the Indian Government officials (personal communication April 2018, Jharkhand) explained that "Traditional knowledge has remained an important source of our decision-making. We develop concepts of environment, forest, and land restoration based on indigenous knowledge emphasizing the symbiotic character of human beings and nature as observed in Sarhul festival (an Adivasi festival in Central India worshipping Sal (Shorea robusta) flowers and trees). There is a mismatch in the continuing indigenous knowledge with the need for moving toward a superpower developed world. For any government, we must lose something from our society to move forward in science, and indigenous knowledge happens to be part of it."

### 1.3.4 Artists – Indigenous Radio

In Colombia, there were several attempts to introduce indigenous radios in indigenous territories. In 2014, a couple of interviews were conducted in the Sierra Nevada region by me during the fieldwork. One such interview (personal communication with an artist in 2014 in Santa Marta) mentioned, "Trying to bring radio to Indigenous Peoples – the Kogi people of Sierra Nevada – is a challenging task. It is not that they do not want to share their indigenous knowledge with the rest of the world, but rather they do not want the central government to introduce modern media at a time when there lacks clarity on what free trade agreement negotiation would lead to. Indigenous knowledge for artists like me might be a way to connect to the secluded world of Kogi people, but for them, it would be stab on their motherland. I think any attempt to simplify indigenous knowledge in any form of communication is risky and poses a threat to otherwise secluded Indigenous Peoples in opening a debate with people who lack interest in indigenous knowledge."

Artivism, a term to explain art used for activism, is gaining popularity. By use of radio, print, podcasts, and audiovisuals it is becoming an important tool for social movements, including by indigenous peoples. Films by and/or about indigenous peoples and their struggle when captured in the voices of the real protogonists it becomes a meaningful discourse in academia as an educational tool but also for civil societies and other stakeholders to demand human rights (Bose, 2023).

### 1.3.5 Academicians (From Higher Education Institutions)

For many academicians researching indigenous peoples, it becomes essential to study what indigenous knowledge means and to define it to publish scientific literature. One of the academicians from Europe (personal communication: Copenhagen, 2018) suggested, "For academicians, we get only a small window to spend time with the indigenous peoples. As such based on the literature review, we often draw analysis to understand indigenous knowledge. Our students doing PhD research add the latest information through their studies. Having a meaningful understanding of what indigenous knowledge is for it to be acknowledged demands time. Most researchers have little say in the way funding is designed and issues about indigenous peoples including indigenous knowledge are the least funded subject in available funding grants. In the given context, we draw our understanding, which is influenced by our Western lens. The bias will exist until an equal number of indigenous scholars are given the opportunity to write through their lens."

## 1.4 MOTIVATION FOR THIS BOOK

As suggested at the beginning of this chapter, understanding indigenous knowledge allows learning and appreciating the ways of living of indigenous peoples their wisdom, knowledge, tradition, and clues of the practices of pacifying the earth and the ecosystem. A comparative qualitative methodology that this book provides is another reason to highlight it as a motivating factor. This book gives a glimpse into the remote landscapes of the Indigenous territories amplifies the voices of the indigenous peoples, and shows the cross-cutting similarities and uniqueness among diverse indigenous communities living in diverse landscapes across continents.

As the author of this book, I recognize capitalizing letters I and P for 'Indigenous People(s)' as a mark of respect and value as political and historical communities. The editorial team of this book, as per publication requirement, has used the words indigenous people(s) without capitalizing the 'I' and 'P' letters for consistency purposes with the style guide.

This book aims to showcase real on-the-ground empirical evidence. In doing so, I maintain an unbiased and ethical approach to avoid creating an ideal image of indigenous peoples. This book chapters maintain the differences within indigenous peoples of being 'stewards' and 'guardians' vis-à-vis those who adopt the mainstream culture of exhausting resources and conflicts. It would be unfair to project only a one-sided version of suggesting that indigenous knowledge and language will continue to persist and resist the 'new paradigm of development.' This book reveals the reality based on empirical case studies about challenges when the culture and indigenous languages disappear. Indigenous peoples are going to be fragmented when the forests and ecosystem's ability to sustain their 'way of living' is exhausted or due to economic or political reasons. The ambition of this book is to amplify existing traditional practices so that youths and the next generations do not dismiss indigenous peoples as 'primitive' and their knowledge systems as 'unscientific' belief systems.

## 1.5 DECOLONIZATION ROADS TO THE WAY FORWARD

Cumulative understanding of indigenous knowledge is the way forward. One of the decolonization roads to the way forward is to work with multi-stakeholders toward a dialogue where indigenous peoples are the decision-makers. This book introduces indigenous knowledge from diverse stakeholder perspectives, including some contrarian views, and sets the tone for the significance of indigenous peoples as stewards of our environment and guardians of some of the remaining rich biodiverse forests in tropical and subtropical regions of the world (Chapters 1 and 2). It also focuses on pertinent issues of tenure and land grabbing, displacing Indigenous communities from ancestral land in Gond Adivasis of Central India, Aymara and Quechua Indigenous Peoples of Peru, and Afro-Colombians from Colombia (Chapter 3). Chapter 4 brings the value of edible foraged/uncultivated forest food for indigenous peoples: the Batwa tribe in Uganda, the Orang Rimba of Indonesia, the Quechua from Bolivia, and the Ladhakhis of North India. Chapter 5 continues to add a perspective of diverse communities with an issue of nomadic pastoralism in drylands in protected forest areas for the Bahima tribe in Uganda and the Maldharis of Western India. Chapter 6 discusses the critical role of factory schools' influence on the indigenous education system and Indigenous languages for the Lai and Warli tribes of India, Huaorani, and Uro indigenous peoples of Ecuador and Bolivia. Chapter 7 reflects on human and wildlife co-existence with evidence from Changpa indigenous peoples, Gurung and Magars in Nepal, Kogi, and Arhuaco of Colombia. Chapter 8 offers reflections on how communication could support bridging science and policy as a way forward. This book is a glimpse into the untapped knowledge that exists in hundreds of indigenous communities that live in diverse landscapes across the globe. It is a document with evidence-based primary data that can find itself in the hands of policymakers, students, philanthropists, and/or practitioners who are still shaping their perspectives and theories.

## REFERENCES

Bartlett, L., and Vavrus, F., 2017. Comparative case studies: An innovative approach. Nordic *Journal of Comparative and International Education (NJCIE)* 1(1):5–17.

Bose, P, 2019. Communicating with 'people' in scientific forestry discourse. Keynote speech at the 25th IUFRO World Congress, Curitiba, Brazil. Accessed via https://www.linkedin.com/pulse/communicating-people-scientific-forestry-discourse-purabi-bose?trk=public_profile_article_view

Bose, P., 2023. Films as human rights: through the lens of Indigenous People's forest food. In B. Wernaart (Ed.), *Applied Human Rights*. Netherlands: Wageningen Academic Publishers, pp. 67–83. https://www.wageningenacademic.com/doi/epdf/10.3920/978-90-8686-943-5_5?role=tab

Gadgil, M., Berkes, F., and Folke, C., 1993. Indigenous knowledge for biodiversity conservation. *Ambio* 22: 151–156.

Gadgil, M., Berkes, F., and Folke, C., 2021. Indigenous knowledge: from local to global. *Ambio* 50: 967–969.

Mistry, J., 2009. Indigenous knowledge. International Encyclopedia of Human Geography.

Nakashima, D.J., and Roue, M., 2002. Indigenous knowledge, peoples, and sustainable practice. In P. Timmerman (Ed.), *Encyclopedia of Global Environmental Change, 5: Social and Economic Dimensions of Global Environmental Change.* Chichester, UK: Wiley, pp 314–324.

Semali, L., and Kincheloe, J., 1999. *What Is Indigenous Knowledge? Voices from the Academy*. New York: Falmer Press.

CHAPTER 2

# 'Nature-Based Knowledge' Aligning Science and Wisdom

**Geographical Location:** Global

## 2.1 INTRODUCTION

Inevitably, in almost all discussions, indigenous knowledge is compared with Western science. This in itself is an unfair comparison. Why cannot the two diverse sets of knowledge systems co-exist and complement each other? Did Western science gain more popularity by minimizing the role of indigenous knowledge? Does all knowledge need to be certified by Higher Education Institutes and reputed international research institutes? The commodification of knowledge – one needs to pay to gain a degree – in Western science is different from non-patented knowledge and wisdom one receives freely from indigenous knowledge. A lot of these wisdoms have now become part of the science, and it continues to contribute except that it has never been patented. Mazzocchi (2006:463) writes that "Western science – which is deeply rooted both in the philosophy of Ancient Greece and the Renaissance – and traditional knowledge systems have developed radically different strategies to create and transmit knowledge and it is exceedingly difficult to analyze one form of knowledge using the criteria of another tradition."

DOI: 10.1201/9781003153061-2

Indigenous knowledge is often synonymously used with terms like traditional knowledge, traditional ecological knowledge, local knowledge, or folk knowledge. The United Nations and World Intellectual Property Organization consider traditional knowledge and local knowledge as types of indigenous knowledge. The term local knowledge could be distinguished by the duration of time they have existed, from millennia to recent decades. In Chapter 1, we discussed indigenous knowledge from a multistakeholder perspective, but indigenous communities themselves may perceive indigenous knowledge differently.

For indigenous peoples, indigenous knowledge is embedded in their day-to-day traditional lifestyle, and it is holistic and connected with ancestors. Some of these practices may be perceived as culturally specific for example Indonesia's Toraja people of southern Sulawesi are known for their elaborate death rituals, which include preserving and exhuming the dead using local preservation techniques. It is regarded as one of the world's most elaborate funeral processes and costs a fortune therefore families keep the deceased person at home. The longer the duration they take for the funeral process the more they have saved money to buy buffalos to be sacrificed. These age-old funeral practices are oral traditions with belief systems that performing the rite and involving everyone in the community for traditional feasts will help bring prosperity. There are similarities in some rituals and symbolism among indigenous peoples across geographical boundaries such as in India's Naga tribe – both in Naga villages and Toraja villages the houses often have carved wooden buffalo horns and real buffalo horns. Though some of these rituals are discontinued in many countries post-colonialism and integration of the community to farm-based livelihoods. The similarities also extend to the clothes woven traditionally by many communities including the Baiga tribes in Central India – they used white unbleached thread – just like the Colombian Indigenous Peoples of Sierra Nevada, the Kogis whose clothing has spiritual beliefs that white represents the Great Mother because it is the purest form of nature.

From the indigenous communities' perspective, any form of harm to nature through misappropriation and abuse of knowledge is unacceptable as it may have spiritual repercussions. This is a reason why sacred groves in many parts of the world, including India and Uganda, are communally protected. These sacred groves act as a pool of conservation of rare fauna and flora, which act as a key source of medicinal herbs. For science, the elements of indigenous knowledge that fit for scientific scrutiny get acknowledged while the rest of 'traditional or indigenous knowledge' that are practices such as sacred groves are seen as 'rules and beliefs' to prevent unsustainable extraction and harvesting of resources from nature. The lack of acknowledging some of these practices, as indigenous communities argue, has led to the depletion of resources including the extinction of rare wildlife species, over-exploitation of sub-soil resources, and deforestation.

One of the major arguments that critics of indigenous knowledge use are that unlike Western Science, indigenous peoples across the globe have relied for millennia on their environment for subsistence and had little or no interest in maximizing production (see Mazzocchi, 2006). The commodification of knowledge or resources was either considered taboo or ill-natured. Extraction of resources, as largely understood in indigenous knowledge is for subsistence purposes and is often done collectively or in consideration of the communities' needs – current and future. Some form of barter system existed (and continues to exist in many indigenous communities) that ensures a sense of interconnection, giving value to social bonding. Just like Western science philosophy, indigenous knowledge has an ethical dimension that limits any forms of exploitation of their nature. The ethical approach of indigenous knowledge is rarely documented as the 'science'; rather, it is categorized as practices and 'rituals' to preserve and maintain the environment and ecosystem. Further, these practices varied between cultures and orders to preserve nature and human symbiosis.

## 2.2 NATURE-BASED KNOWLEDGE: ANALYTICAL FRAMEWORK

Using the indigenous knowledge paradigm as the framework, this chapter argues that following colonization and missionary excursions in many parts of the sub-tropical and tropical countries, modern or Western science was introduced. With this came notions of cultural colonialism, which tended to depreciate many forms of indigenous knowledge and co-existence with nature and challenged its existence and validity. The challenge was biased because the criteria used to assess the indigenous knowledge was from the perspective and set of rules that Western science has identified.

Even before the advent of modern-day technology to forecast meteorological phenomena, indigenous knowledge shared information to cope with climate challenges. This form of nature-based knowledge to be acknowledged needs a major change in the way scientists think about what 'science' means and the way scientific output is labeled.

Many scientists are tuned in to trust only evidence that is produced from thorough Western science-based guidelines that are quantitatively and qualitatively analyzed data that proves the validity, which is difficult to change due to colonial legacies (see Schipper et al., 2021; Ankrah et al., 2021). Much of the current-day climate research has excluded indigenous knowledge even though indigenous communities for millennia have used it to adapt to the climate crisis.

For an inclusive approach to indigenous knowledge, various efforts have been made to provide some legal protection for indigenous peoples in the form of indigenous intellectual property – predominantly based on Western concepts of intellectual property law. Indigenous intellectual property rights aim to include cultural knowledge, traditional medicines, arts, and literature including historical sites, textiles, music, songs, dance, and ceremonies. There have been various Declarations across the globe such as Kari-Oca Declaration in Brazil (1992), Lakota Declaration (1993), and Kimberley Declaration (2002), and the most recent and important is the United Nations Declaration on the Rights of Indigenous Peoples (UNDRIP). It is a legally non-binding resolution that upholds the individual and collective rights of indigenous peoples. A few of the UNDRIP clauses have highlighted the intellectual property rights of indigenous peoples. The Declaration (United Nations, 2007) recognizes: the urgent need to respect and promote the inherent rights of indigenous peoples, which derive from their political, economic, and social structures and from their cultures, spiritual traditions, histories, and philosophies.

***Article 11*** indigenous peoples have the right to practice and revitalize their cultural traditions and customs. States shall provide redress through effective mechanisms, which may include restitution, developed in conjunction with indigenous peoples, with respect to their cultural, intellectual, religious, and spiritual property taken without their free, prior and informed consent or in violation of their laws, traditions, and customs.

***Article 24*** indigenous peoples have the right to their traditional medicines and to maintain their health practices, including the conservation of their vital medicinal plants, animals, and minerals.

***Article 31*** indigenous peoples have the right to maintain, control, protect, and develop their cultural heritage, traditional knowledge, and

> traditional cultural expressions, as well as the manifestations of their sciences, technologies, and cultures, including human and genetic resources, seeds, medicines, knowledge of the properties of fauna and flora, oral traditions, literature, designs, sports and traditional games and visual and performing arts.

Many indigenous communities have been producing food and drinks using non-timber forest products such as flowers, fruits, and roots. The knowledge they gained through oral tradition is practiced by the communities collectively though their knowledge has not yet been patented. Private companies interested in marketing these products for commercial purposes are being monitored due to these Declarations; otherwise, this knowledge would have been appropriated. Free, Prior, and Informed Consent is another important tool for indigenous communities when private and government investors attempt to develop products using indigenous knowledge or develop projects that may affect their lands, territories, and resources such as rivers and forests. indigenous peoples' right to give or withhold consent on any activities is related to their right to self-determination.

Indigenous knowledge as nature-based knowledge, as this book proposes, takes the analytical approach that it needs to be understood as holistic rather than fragmented compartmentalized lens. Western science is about being objective based on 'facts' or quantitative data while indigenous knowledge is qualitative with emotions and subjective (Mazzocchi, 2006). As Freeman (1992) explains, unlike Western science, indigenous knowledge is made up of multidimensional in a way that all other elements are crisscrossed in a complex web of interactions. For the purpose of this book, the framework of nature-based knowledge reflects that indigenous knowledge to be acknowledged demands accepting the diversity of two sets of knowledge that might be parallel, but it does have a merging path when explored further. All actors on this earth need to realize and be aware that the knowledge we hold is the entry point as well as the base for the perspective we draw to understand indigenous knowledge. This is even true for the analysis presented in this book – a perspective of indigenous peoples themselves is documented as narrated to me and analyzed by me using the lens of indigenous peoples. It is this consciousness that determines where we belong and where we want to belong – a path to both the world of Western science and indigenous knowledge.

## 2.3 INDIGENOUS PEOPLES LIVING IN AND AROUND FORESTS: UNITY AND DIVERSITY

Forests are the most valuable resource on this planet. It is estimated that over 2 billion people's livelihoods across the globe are supported by forests and forest resources. Moreover, it is home to over half of the world's terrestrial species of animals, plants, and insects. Almost one-fourth of the world's land surface is considered to be indigenous territories. The tropical forests of the Congo Basin are itself home to about one million indigenous people. Deforestation in the 'Lungs of Africa' is competing with the Amazon forests as one of the remaining important carbon sinks in the world. Roads and other improved infrastructure access to forests in the Congo Basin, but also in many forested landscapes has increased the deforestation rate. It is estimated that in 2022 Brazil accounted for 43% of felled trees, and Congo Basin about 13% of forest destroyed. Many indigenous peoples are displaced, and they migrate to urban areas for work due to the lack of their traditional forest-based livelihoods. Many governments fail to recognize them as indigenous peoples (Bose, 2012), often leaving these ethnic groups vulnerable to a lack of forest tenure rights and access to resources.

The indigenous peoples are regarded as 'guardians of the forest' and 'stewards of the land' due to their ability to manage the forest using indigenous knowledge. However, they pay a heavy price in return individually and collectively.

### 2.3.1 Environmental Defenders

Radwin (2023) reports that at least 177 people were killed in 2022 for defending the environment, with a fifth of killings taking place in the Amazon rainforest. Colombia was the deadliest country. Overall, indigenous communities made up about 35% of all murders, although they represent only 5% of the world's population. Between 2012 and 2022, according to Global Witness, about 1910 environmental defenders were killed. A risk and price that indigenous peoples pay for being the guardians of the forests. Environmental defenders are ordinary people who are often subjected to harassment, violent attacks, and death threats before they are murdered. Indigenous environmental defenders face more risk due to remote locations with minimum or no protection for their safety. Not all environmental defenders are murdered, but rather, they are imprisoned on fabricated and false charges, making it difficult for them to defend. Often, they are subjected to smear campaigns in local media alienating them from the community as someone who is corrupt or is against people's development. In many instances, institutional environmental defenders face inquiries about receiving overseas funding and are portrayed as using funding for illegal activities such as anti-national activities. It distracts individuals and institutions from their main fight to protect the environment and rather unsuccessfully spend their resources to clean their reputations.

### 2.3.2 Forest Tenure

Forest tenure is broadly defined as the customary or statutory rights about access, use, manage, control, and/or transfer of forest lands and resources, making the primary decisions about forest resources. It may or may not include ownership and tenancy rights. The complexity of forest tenure is magnified by the vague or unclear definitions of the term 'forest' and 'forest-dependent or forest-dwelling communities.' Moreover, political conditions and historical and cultural factors influence the way forest tenure is implemented because forest land particularly sub-soil resources and forest resources such as non-timber forest produce are seen as two different aspects of forest policy. In other words, indigenous knowledge has been rarely integrated or discussed as a core element of determining what is tenure security for indigenous peoples across the globe. With a lack of political recognition of indigenous knowledge, any form of decentralization of resources including power will fail as long as tenure security lacks respect and acknowledgment by a political and economic system. Recognition of forest tenure becomes more critical for indigenous women and children who suffer the most from the Western science-promoted forest conservation and land-use change. Forest tenure is a Western concept, and largely written by Western scholars from their perspective – for indigenous peoples' perspective to be inclusive the concept of tenure would need to be redefined. In 2023, India's Central province, Chhattisgarh, granted habitat rights to the Baiga tribe. Habitat rights in India recognize the community of their customary territory of habitation, socio-cultural practices, economic and livelihood means, intellectual knowledge of biodiversity and ecology, traditional knowledge of the use of natural resources, as well as protection and conservation of their natural and cultural heritage (see Naidu, 2023). Though the habitat rights are limited only to certain categories of tribes – Particularly Vulnerable Tribal Groups – there is a huge potential in many indigenous territories to implement decentralized habitat rights in lieu of limited forest tenure governance.

### 2.3.3 Climate Adaptation

Filho et al. (2022) explain that it took the sixth assessment of the Intergovernmental Panel on Climate Change to articulate complementarity between indigenous knowledge and scientific knowledge, unlike the previous assessment report that excluded indigenous knowledge because of a lack of empirical evidence. Further, they mention that the thematic prioritization of funding for climate change research in Africa shows that institutions in Europe and North America (the Global North) received over 75% of funding for climate research while African institutions received only 14% (ibid). This discrepancy is combined with critics of climate change research that a majority of scholars and publications on indigenous peoples are by the Global North with few or no involvement of indigenous scholars and communities. Filho et al. (2022) argue that 'climate research needs to be more flexible with regard to its approach and that it should not only be grounded in the Western perspective of "science."' The United Nations Sustainable Development Goals have emphasized the need to acknowledge indigenous knowledge from diverse communities in adapting to climate change such as nature-based knowledge about preventing forest fires to observing the environment – winds, clouds, and the movement of wildlife to forecast hurricanes or tsunamis, and even knowledge about herbal medicines in the wake of a pandemic.

To summarize, this chapter highlights that the term indigenous knowledge has no single definition but has many common understandings from the Western science perspective. Moreover, the distinction between indigenous knowledge and Western science is a risky comparison and can derail the complementarity of two different types of knowledge systems. Lastly, it highlights that though indigenous knowledge has been acknowledged in various international declarations the pathway to accomplish it in practice remains far from complete. Forest and environment remain an important source of preserving indigenous knowledge and therefore all development activities including tenure rights, forest and landscape restoration, forest policy, and protection of environmental defenders must acknowledge indigenous knowledge by including their voices in decision-making.

## REFERENCES

Ankrah, D.A., Kwapong, N.A., and Boateng, S.D., 2021. Indigenous knowledge and science-based predictors reliability and its implication for climate adaptation in Ghana. *African Journal of Science, Technology, Innovation and Development* 14(4): 1007–1019.

Bose, P., 2012. 'Forest governmentality': A genealogy of subject-making of forest-dependent 'scheduled tribes' in India. *Land Use Policy* 29(3): 664–673.

Filho, W.L., Wolf, F., et al., 2022. Is indigenous knowledge serving climate adaptation? Evidence from various African regions. *Development Policy Review*. https://onlinelibrary.wiley.com/doi/epdf/10.1111/dpr.12664

Freeman, M.M.R., 1992. The nature and utility of traditional ecological knowledge. *Northern Perspective* 20: 7–12.

Mazzocchi, F., 2006. Western science and traditional knowledge. Despite their variations, different forms of knowledge can learn from each other. *EMBO Reports* 7(5): 463–466. doi: 10.1038/sj.embor.7400693

Naidu, J.S., 2023. Baiga tribal group gets habitat rights in Chhattisgarh: What this means, significance. *Indian Express Report*. Accessed via https://indianexpress.com/article/explained/explained-politics/baiga-pvtg-habitat-rights-chhattisgarh-significance-8976933/

Radwin, M., 2023. Latin America most dangerous place for environmental defenders, report says. *Mongabay*. https://news.mongabay.com/2023/09/latin-america-most-dangerous-place-for-environmental-defenders-report-says/

Schipper, E.L.F., Ensor, J., Mukherji, A., Mirzabaev, A., Fraser, A., Harvey, B., Totin, E., Garschagen, M., Pathak, M., Antwi-Agyei, P., Tanner, T., and Shawoo, Z., 2021. Equity in climate scholarship: A manifesto for action. *Climate and Development* 1310: 853–856.

United Nations, 2007. Declaration on the Rights of Indigenous Peoples accessed via https://www.un.org/development/desa/indigenouspeoples/wp-content/uploads/sites/19/2018/11/UNDRIP_E_web.pdf

CHAPTER 3

# Extractive Industries Mining Way in Indigenous and Local Communities

**Geographical Location:** India, South Asia; and Colombia and Peru, Latin America

**Indigenous Peoples:** Afro-Colombians, indigenous peoples of Valle de Cauca; Peruvian Amazonia indigenous peoples including Quechua and Aymara; India's Chhattisgarh Adivasi Gond tribe

## 3.1 INTRODUCTION

In modern living, mining, minerals, and metals are critical aspects of economic development for all countries. Both developed and developing countries are dependent on mining for electricity and essential materials needed to build our homes, hospitals, and transportation systems. The current debate is shifting from questioning whether mining is important to how mining could better contribute to sustainable development. Mining companies are forced to re-think ways of their business to collaborate with diverse supply chains to reduce waste, reuse, and recycle. In practice, there are several reports and claims that the mining industry can be devastating to the forests and natural resources, and in particular


DOI: 10.1201/9781003153061-3

to the individuals and collectives who are defending their land. According to the Responsible Mining Foundation report (RMF, 2020), many mining companies are failing to demonstrate human rights commitment, which has led to an increased number of threats and killings of environmental defenders about mining operations and extractive projects. The World Bank's Executive Summary suggests creating a supportive enabling environment for forest-smart mining principles, including secure tenure over forests and forest resources that safeguards and respects indigenous and other customary rights, to support long-term forest stewardship and sustainable use (World Bank, 2017). Yet, mining is an important driver of deforestation (Bose, 2023; Siqueira-Gay et al., 2020).

Mineral extraction from forests brings a plethora of other interventions such as road infrastructure which further opens up the door to deforestation, loss of biodiversity, and human rights abuse. These cumulative impacts – defined as "a change in the environment caused by multiple interactions among human activities and natural processes that accumulate across space and time" (CCME, 2014) – of mining are important to understand how forest-dwelling communities' indigenous knowledge (IK) is affected. Mines, although estimated to be 1% of Earth's land surface their impact on forests and land use change is beyond any form of estimation – fragmentation of social, economic, biodiversity, cultural, and traditional forest-related knowledge. The United Nations (2004: section I.3) suggests "Traditional forest-related knowledge is a cumulative body of knowledge, practice, and belief, handed down through generations by cultural transmission and evolving by adaptive processes, and the relationships of living beings (including humans) with one another and with their forest environment." Mining is a lucrative business, and often money is invested through illegal activities for example Narcos in the Amazonian region or the Revolutionary Armed Forces of Colombia (FARC) in Colombia, or local timber mafias in India. Villar and Schaeffer (2019) in their study finding suggest that in the Afro-Colombian Pacific States, unregulated mining is the main driver of deforestation, but the cumulative impact is the spread of malaria and other insect-borne diseases such as babesiosis that is ignored by the informal job market run by ex-guerrillas, organized crime groups and corrupt officials that sustain the black markets ignoring the indigenous peoples and local communities.

In this chapter, I examine aspects of the impacts of mining in forested landscapes on their social, cultural, practices and beliefs, and traditional governance of forest and its resources, which are considered together as 'indigenous knowledge.' To do so, I use three case studies: two on gold mining from Latin America – Afro-Colombian region, Colombia, and Peruvian Amazon, Peru, and one on coal mining from South Asia – Chhattisgarh. After this introduction, the remaining chapter is divided into four parts. Part 2 is about the case study data collection method and ethics are explained; parts 3 and 4 are three case studies and discussion, respectively, and the final part is about drawing conclusions and policy recommendations.

## 3.2 ETHICS, METHODS, AND DATA COLLECTION

Mining is a sensitive topic and local communities affected are vulnerable. The basis of this study is the ethical dimension that shaped the methods used and techniques for data collection to ensure the privacy of the participants. One of the ways to gain trust with the communities was to approach them with the help of local partners, with whom they have long-term collaboration. In addition, being honest about data access, using pseudonyms, informed consent, being respectful of the research site and people, reciprocity, maintaining confidentiality, and being transparent about the use of the knowledge without manipulation or falsification. As an experienced ethnographic researcher, I maintained ethical considerations in my empirical study. Knowledge of the local language is crucial when collecting data, and I learned Spanish and a bit of Quechua before embarking on my fieldwork in Peru and Colombia. I had a field assistant who supported me in in-depth interviews. For Chhattisgarh, everybody spoke Hindi, a language I am fluent in, but to ensure knowledge is lost in translation, our local guide knew Gond's indigenous language and fluent Hindi.

In India and Peru, the data collection was challenging on some days due to informal inquiries by the mining company agents about the research purpose and outcome. In Madre de Rios, the community was protective of me to ensure that there was no harassment faced by me during the fieldwork. They assigned two male youths and a woman to accompany me during interviews of the household. We had to cross rivers by boat and walk in forest areas without access to a phone connection for an hour to reach the community. In Popayan, I was accompanied by a research assistant, a middle-aged man, who was Colombia's ex-FARC member. For fieldwork in India, despite having prerequisite permission to do research, we were followed by a group of men – who were identified by the community members as local unelected politicians and/or sub-contractors of a mining company – during one of the fieldwork days when we visited the mining site.

Typical procedure before data collection was sharing information about the researcher and her affiliation, the purpose of the study, the type of participants, potential risks, the option to withdraw at any time, and oral consent. The oral consent was mutually agreed upon – individual as well as collective consent – to avoid any identification with their identity with written consent. Except in India, photos and videography were discouraged to avoid attracting the attention of miners. We filmed only in locations where filming was allowed with the consent of local communities. One such film, Kabza, is produced by me as a self-funded initiative, Landing Together Films (https://www.youtube.com/watch?v=KpqEZE1EefY).

This chapter uses qualitative case studies, the method for empirical data was in-depth interviews and focus group discussions with diverse stakeholders in the study areas. In total, 47 individuals were interviewed (Peru: 14; Colombia: 16; and India: 17) at least 30% were women interviewees in each study location. In addition, at least two to three focus group discussions were organized in each of the case study areas. In total, 12 focus groups with 195 stakeholders participated in this study. Peru and India had the most diverse multistakeholder actors – often involving diverse local leaders, elders, and youths – that the community decided to invite (see Table 3.1). Due to armed conflicts, the Colombian case study had the least number of stakeholders, but some individuals were allegedly former members of armed conflict groups without self-identifying as one.

In each case study area, interviews were conducted with the support of a research assistant or a representative of a local institute familiar with the local language and dialect and English, and often, the semi-structured interviews were about 50–60 minutes. No financial compensation was given to the interviewees because the study was in collaboration with the local partner institutes and the institute of my affiliation. The local institute

TABLE 3.1 Representation of the Individual Interviewees in the Case Study Areas

| | **Peru** | **Colombia** | **India** |
|---|---|---|---|
| **Focus Groups** | 4 ($n = 64$) | 3 ($n = 19$) | 5 ($n = 112$) |
| **Individual Interviews** | **Peru ($n = 14$)** | **Colombia ($n = 16$)** | **India ($n = 17$)** |
| Cooperative or village local institution head | 1 | 2 | 2 |
| Women leader/representative | 3 | 1 | 6 |
| Youths | 6 | 2 | 5 |
| Elders/indigenous leaders | 2 | 6 | 3 |
| Local politician | 1 | - | - |
| Community/villagers | 1 | 5 | 1 |

in all three locations works with the communities on the issue of mining providing them legal, social, and economic support, as well as training to enhance their traditional knowledge skills related to forest and land use.

## 3.3 CASE STUDIES

### 3.3.1 Peru | Madre de Dios

In 2007, Peru voted in favor of the United Nations Declaration on the Rights of Indigenous Peoples and ratified ILO Convention 169. A country with over four million indigenous peoples, they are diverse, made of 55 different groups who speak 47 different languages. Many do not speak Spanish or Quechua except their native language. About 5% of this population belongs to Amazonian indigenous peoples. Peru is considered to be the largest gold producer in Latin America. After Brazil, Peru has the second-biggest share of the Amazon Rainforest but with an average loss of more than 326,000 acres of forest per year since 2001 (Brown, 2023).

The Ministry of Energy and Mines, along with various other government agencies, regulates the industry. To obtain a concession, an environmental impact assessment (EIA) is a prerequisite. However, EIA takes a backseat given that the demand for the precious metal is high making illegal activities a boom. Almost 25% of gold mining is estimated to be illegal and often carried out in Peru's prohibited land. The impact of the gold rush has left negative environmental and social consequences, not just making it one of the main drivers of deforestation but also high human rights abuse in the Peruvian Amazon.

The case study area where I did fieldwork is known in Peru as the 'gold mining corridor' and lies in Madre de Dios in the Peruvian Amazon, in the country's southeast region (see Map 3.1). The fieldwork was carried out in illegal gold mining in 2014. According to the indigenous communities, thousands of illegal mining sites are functional without any mining concessions from the government. These illegal activities are difficult to control due to limited monitoring by the government, and a lack of formal collective forest tenure rights by the local communities. In the study area, I visited two gold mines that were illegal and had deforested over a hundred hectares. One of the local environmental defenders was killed a few months before my fieldwork, who was protesting against illegal logging and pollution of the river. This indigenous community traditionally has been fishing and practiced hunting-gathering as livelihood subsistence in the Amazon. "Across the region (Madre de Dios), I have seen illegal logging and illegal mining by outsiders that has left us (indigenous peoples)

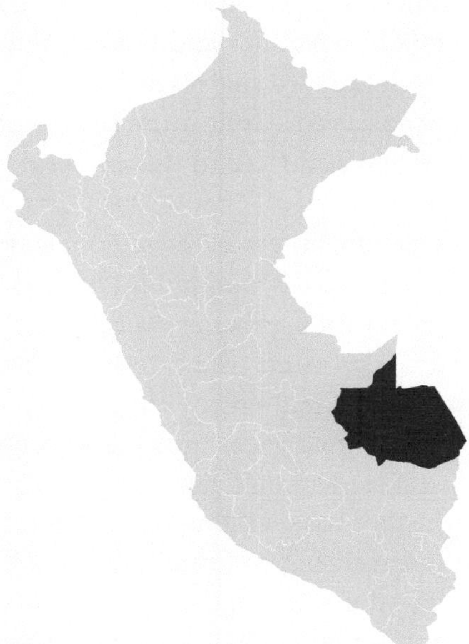

MAP 3.1 Map of Peru.

living in the area devoid of any livelihood resources and destroying our traditional way of living. They (gold miners) hire non-indigenous people and treat them like slaves. In the last two years, even some of my community youths have in desperation taken up labor work in gold mines and they are facing health issues due to pollution. Our generations of conservation have gone waste; why did our parents and grandparent had remained poor and conserved the forest and subsoil minerals?" says one of the local community members who is suffering from mercury contamination.

During the focus group discussions, the community discussed their identity as Amazonian indigenous peoples, which is lost due to assimilation with many outsiders settling in the region and all claiming to be local communities. The land use change due to illegal gold mines has left rivers and soil polluted with mercury and the community has no choice but to continue fishing as it is an important part of the diet. The IK to conserve nature – including when to fish, how to fish sustainably, and which variety of fish is beneficial for pregnant women or for arthritis in elders – holds no value any longer. An indigenous woman during the interview mentioned,

"We know over 200 different species of fish just in this region. Each species has its name in our language; we know all the details about them as oral knowledge. We talk about fishes and rivers like they are our family. IK does not work in isolation. To maintain fish in our rivers, the entire ecosystem should be conserved. Gold mine-related mining pollution is fragmenting our ecosystem and IK, and this mineral has no meaning for our community and nature."

The community, like many others in the region, is struggling with mining pollution. The focus group discussion highlighted that the communities considered the conservation effort as a piecemeal solution to mining rather than a cumulative approach. The external conservation, according to the community, was concerned only with 'plantation' in the deforested area while the mercury-polluted water in the Amazon was not dealt with. One of the local political leaders (not elected at the time of the interview) mentioned, "One of the most challenging aspects of mining is corruption. It is so rampant and indirect that communities are unable to deal with it. Many authorities involved are taking bribes from mining companies to 'allow' them for illegal mining. How can local communities protect their forests and land when those elected are stealing and selling it away for some cash? Sometimes, they even bribe one or two local community youths or buy them prostitutes and alcohol to spread fake news that mining will bring development to the community. These issues are not addressed by any institutions because everyone fears them (mining companies/agents)."

The corruption is subtle and none of the stakeholders other than the community were ready to discuss the issue, and/or have denied the existence of corruption. Evidence to prove that ecosystems are damaged due to mining companies is counterargued by the authorities claiming it is the community who is involved in artisanal mining leading to the deterioration of the flora and fauna. One of the youths in the study area explained, "A few community members are influenced by the mining companies by bribing them – a monthly payment or offering them prostitutes to spread fake news that 1000 acres of mining land are due to artisanal mining by the community. We are set up against each other leading to fragmentation of traditional indigenous community bonding and knowledge sharing." At the time of data collection, the respondents explained that earlier, their dependence on forest resources through hunting, fishing, and using some timber for the household was met, but now with mercury poisoning and heavy use of machinery mining has radically changed the landscape of their region leading to deforestation and loss of biodiversity.

### 3.3.2 Colombia | Popayan

Colombia's Popayan is famous for its architecture and for being Latin America's first UNESCO-designated city of gastronomy in the Department of Cauca. In 2013, at the time of fieldwork data collection, Cauca was a FARC stronghold. FARC or the Revolutionary Armed Forces of Colombia – People's Army was founded in 1966 by peasant self-defense groups and, over the years, has adopted unconventional methods, including kidnap, illegal drugs, and illegal mining. At the time of my fieldwork, the president of Colombia and FARC were negotiating a peace deal. The signing of a ceasefire accord happened in 2016.

The Department of Cauca has a long history of gold mining production, a reason why Popayan became Cauca's capital (OECD, 2017). Since the 1550s, indigenous and Afro-Colombian communities were used as cheap labor, making them dependent on artisanal mining for subsistence (*ibid*).[1] The study site (see Map 3.2) was on the outskirts of Popayan, bordering another municipality. The indigenous communities interviewed are divided in opinion about mining. I decided to do this field study on gold mining in FARC-dominated territory and an Afro-Colombian community when I was in Popayan supervising my intern, a Dutch Master's student. There, I met one of the Afro-Colombian women who narrated how it was unsafe for her to be in her village. She explained how IK is fading away due to a myriad of problems – from drugs to alcoholism, to kidnap and crime among her community. No one was addressing this issue and she wanted me to do a research study to highlight the challenges of women and the community as a whole.

Women in the study area raise their voices regarding their rights to land and natural resources, as well as about crime, health, and disparity. The highlights of the focus group discussion were as follows: We are rural Afro-Colombian women and indigenous women in a country where white women are only seen as a beauty product of the cosmetic industry to enlarge their breasts or plastic surgery. We are motivated to defend our IK by seeking to vindicate our land rights, but so-called multistakeholders deny our rights because we are women, we are black, we are indigenous, and we talk about IK that these stakeholders do not know about! We, women, face the brunt of every sword.

"Historically, we Afro-Colombians are reduced to no identity with no traditional knowledge. Ours is a case of no return to the traditional way of living because we do not know if we had one. Unfortunately, like us, the indigenous communities are going through a similar situation, today we

MAP 3.2 Map of Colombia.

are all becoming artisanal gold mining laborers like us. Indigenous and Afro-Colombian community land is used for small-scale mining activities, we get cheap labour wages, and the mineral goes to the wealthy. Our collective lands are destroyed by extractive activities. If we protest, we would be killed," said one of the leaders from indigenous and Afro-Colombian communities. In the focus group discussion, some community members strongly opposed the presence of miners and entrepreneurs who exploit their land and resources, while some members accepted miners in their territories as no choice. According to the discussion, some of the miners, all men, are former FARC or use the FARC identity to threaten those in the community who oppose mining.

The study community is part of an artisanal miner association making them decision-makers of the cooperative. The cooperative's main aim was to support the local communities for mining titles. However, the lack of resources meant the communities had to bring in intermediary agents who in turn controlled their land use, often illegal coca farms. The community has been traditionally secluded due to conflict, and weak state presence has further provided a ground for illegal gold mining. According to one of the indigenous leaders of the community, "We have received threats from non-state armed groups, some of our children are kidnapped, our women faced abuse, all because we rejected mining operations in our traditional land. We heard in meetings by civil societies about the prior consultation process, but that does not exist in practice here; I received a death threat."

'Fear' has been a constant term that has dominated this case study's interviews and discussions. Many community members expressed helplessness and expressed, "Narco group has money and power, and they would kidnap our youths or frame them in crimes they did not commit. We fear they can misuse power and political connection; they can bribe authorities who are corrupt therefore many of us out of fear maintain amicable relationships and adhere to their demands related to mining and land-use change," said one of the members participating in the focus group discussion. One of the municipal authorities, whom we interviewed with the promise to keep identity anonymous, estimated that the region's gold production through artisanal mining is high at 70% and almost half of it takes place in the indigenous territories and land and forest reserves managed by Afro-Colombians. However, according to this authority, not all informal mining is illegal, but it may have been funded by criminals who take over the land after mining. Illegal gold mining is so lucrative that the government pays little or no attention to IK and communities such as Afro-Colombians and intentionally fails to formalize artisanal and small-scale mining (El Pais, 2015). "The artisanal miners are vulnerable and lack connections selling their products at 'fair market value' and therefore they have no choice but to fall back on the criminal groups and illegal armed forces and get exploited that leaves them without their land and identity," he argues.

In the nearby community, there have been protests while I was doing fieldwork. Upon my return to the office in Cali, I was informed that several artisanal miners were killed by paramilitary in the neighboring community of this study area. One of the champions of the protesters was a woman, Francia Márquez, whom I was unable to meet during my fieldwork; she was a famous activist and lawyer in the region fighting against illegal gold mining, particularly in her Afro-Colombian community, La Toma. According to the local authority at the time of fieldwork, he said, "Francia has received several death threats, but she continues her activism on how the conflict has disproportionately affected Afro-Colombians, particularly black women and ethnic minority groups." In 2023, at the time of this book going to publishing, Francia Márquez is serving as Vice President of Colombia and was also appointed as Minister for Equality and Equal Opportunity of Colombia.

### 3.3.3 India | Chhattisgarh

India's Chhattisgarh is a prime example of a state with a 'resource curse.' It was granted statehood in late 2000 with a population of about 30 million but with the third largest forest cover in the country, with over 40% of the state covered by rich biodiverse forests. The state (see Map 3.3) is also diverse with indigenous communities with about 42 different ethnic tribes and the majority being the Gond tribe. Forests and indigenous communities are intertwined in a way that indigenous communities, known as Adivasis, are forest dwellers, traditional hunter-gatherers, and guardians of the forests.

The dense green biodiverse forest area is threatened by the coal mines that have engulfed the forests, as well as by land and the human rights of the Gond people. In the study area, one of the interviewees pointed out, "They blame us for deforestation, but the reality is that they cut the forest, take the timber, and leave us without any forest food that we collected from those trees. Then, they sell our traditional forest land rights to mining companies to extract coal with a false promise to give us employment, but we get cheated as outsiders are brought in as laborers and they are settled in our land. All this only fragment us. Many of our community members have left their traditional homes and migrated to cities for daily wage labour work."

The water wastage from the coal companies is released into the water bodies or pumped underground, only polluting the drinking water of the communities. In this context, women leaders played an important role in

MAP 3.3 Map of India.

protesting against the mining companies and demanding for clean water supply. "How can we allow our fish and farm crops to die, it is our land and they (miners) come here to extract our mineral resources and then pollute our water. Is this even fair?" asks one of the Gond women leaders who took against the mining company. Due to mining pollution, a majority of the local indigenous farmers have lost their crops and collective rights over ponds for fishing. Kodo millet is considered by the indigenous communities as a life-saving medicine and is cultivated in a traditional method.

The community's IK related to practicing traditional farming but also using those indigenous millets has been negatively affected due to mining activities in their forests and land. An elderly couple, who has lived in the study area said, "We were born here, and we lived with wild animals;

we were taught to respect wild animals, trees, but also air, water, soil, and everything that is underneath the soil. We consider it sacred, and our mountains and forests are our religion. In Gond, our last name is often referred to as a wild animal, like my surname means turtle. That's how we begin respecting and learning not to extract but to conserve." The Gond indigenous peoples' life is interlinked with the biodiversity that exists around them, and their IK is in their culture and ways of living. Their culture of food habits is based on the knowledge of what kind of wild edible food is foraged and in what season. Their houses are environment friendly, made of mud, stones, and timber from their common land, which is designed in a way that helps them to keep the temperature lower during hot weather, and in the rainy season, it does not trap moisture. The story of their lives and encounters with wildlife is painted on the outdoor wall of their house, which narrates the story of their co-existence with nature, and one of the morals in all these stories is how to protect the nature earth and not exploit more than what is needed for basic way of living.

The focus group discussion highlighted the following: the government, and the mining industry if have to sell the idea of mining, have to do it differently. They cannot ignore our traditional knowledge of taking care of biodiversity conservation, and management of forest resources. One of the concerns that was expressed during the discussion was about 'slow poisoning' due to coal mining dust and toxic waste released contaminating underground water. An increasing number of health cases were reported, particularly among the elderly and children. The community's traditional medical knowledge was not helpful because the symptoms were new, which many are certain is because of the mining-related pollution. The healthcare center refuses to identify a large number of health problems that the communities in the study area were experiencing that could be linked to the coal industry. The other concern that was raised during the focus group discussion is that the community, through a local non-profit organization, was made aware of Free, Prior, and Informed Consent (FPIC), as well as about the EIA. They realized what their collective land rights were, how the EIA was not conducted systematically, and that FPIC was never implemented. This is a reason why the community began protesting against the coal mining company and also demanding recognition of their IK. The protests were non-violent against the company, and the community was never compensated for the damage to their property or

The customary rules continue to be in practice by the community members, particularly elders. The establishment of the Iron ore company in the community's common land further aggravated the situation, leading to violent protests. The authorities have evidence that the community gave consent and accepted compensation, but the community members claim that the consent was not collective but rather bribed to one of the estranged community members. The violence has led to the arrest of several members of the communities who have protested against the mining companies. "We are scared and fear for our lives. We, women, can no longer protest because they attack our men. They know by doing so, we are silenced," said the elected village institution (Panchayat) women leader.

## 3.4 DISCUSSION AND CONCLUSIONS

This chapter examined how extractive industries without having FPIC destroy the ecosystem that has been preserved for centuries by the communities. From the three case studies, Colombia, Peru, and India, it is evident that the sub-soil rights with the government often fragment the communities of their land, consequently loss of IK.

The above three cases also demonstrate how mining is negatively influencing the local communities' practice of IK, but also it brings fear and new forms of 'knowledge' such as corruption and bribery and evokes anger and violence.

**Indigenous peoples exercise a variety of strategies through resistance and negotiations.**

This chapter provides ethnographic documentation of strategies used by indigenous peoples to claim their land and forest rights that are being grabbed by mining companies. It illustrates that resistance or dependence on mining by indigenous peoples in gold or coal mining is based on various factors that are not limited to their traditional forest and land rights. It is evident from all three cases, indigenous participation in official decision-making did not exist at the time of the fieldwork. It also illustrates that the multistakeholder negotiations are either non-existent or, if at all, it is top-down. From the three cases, findings highlight that the relation between different stakeholders matters – whether local communities decide to c with armed groups as in the case of Colombia, or whether protest as in Peru and India's case studies.

**Indigenous knowledge is significantly constrained by the legal, political, and economic state that prioritizes mining.**

To refer to Scott's (2020:293) findings in relation to extractive contracting and the struggle for control of indigenous lands, "Consent by contract is a technique of the contemporary settler state to normalize its own tenuous authority over lands and resources in the face of increasingly convincing counter-claims." To conclude, a few key policy recommendations based on this study, I would recommend that FPIC should be mandatory instead of voluntary guidelines to be implemented by the state. There is a need to study the nexus of illegal miners and government authorities that very little research has so far addressed due to the danger associated with it (see Asori et al., 2023). The underlying connections between the armed groups, illegal miners, drug dealers, gold traders, and paramilitary groups cannot be ignored in studying IK and its impact on the traditional livelihoods of the local communities. The final recommendation is to include indigenous peoples, Afro-Colombians, and women in the decision-making political process, which is critical in bringing positive change.

## NOTE

1 According to OECD 2017 (see page 20) "In 1559 a letter from the governor of Popayan mentioned that more than 6,000 indigenous and 300 Afro-Colombian people were working 250 days a year in gold exploitation. In 1567, another letter added that 'It is a wonderful thing that metals in this Province are discovered every day. If it were possible to continue exploiting them, I believe there would not be anything more important in the Indies' (Es cosa maravillosa los metales que en esta Gobernación se descubren cada día, que si hubiese posible para poderlos seguir, creo que no habría cosa más importante en las Indias"

## REFERENCES

Asori, M., et al., 2023. Is illegal mining socio-politically entrenched? An opinion piece of the interaction between formal politics and chief dominance in mineral governance, and its influence on fighting Galamsey in Ghana. *GeoJournal* 88(2): 1953–1963.

Bose, P., 2023. Equitable land-use policy? Indigenous people's resistance to mining-induced deforestation. *Land Use Policy* 129: 106648.

Brown, K., 2023. Forest conservation efforts in Peru are failing across the board, study says. An online news article at Mongabay accessed on 2nd June via https://news.mongabay.com/2023/05/forest-conservation-efforts-in-peru-are-failing-across-the-board-study-says/

CCME, 2014. Canadian Council of Ministers of the Environment Canada-wide definitions and principles for cumulative effects, accessed on March 2024 via https://ccme.ca/en/res/cedefinitionsandprinciples1.0e.pdf

El País, 2015, 13 March. Los tentáculos de la minería ilegal en Cauca, accessed on January 2016 via www.elpais.com.co/elpais/judicial/noticias/tentaculos-mineria-ilegal-cauca

OECD, 2017. Due diligence in Colombia's gold supply chain. Gold mining in South West Colombia. https://mneguidelines.oecd.org/Colombia-gold-supply-chain-overview.pdf

RMF, 2020. Summary of responsible mining foundation report. Accessed online on 1st June 2023 via https://2022.responsibleminingindex.org/en

Scott, D., 2020. Extraction contracting: The struggle for control of Indigenous lands. *South Atlantic Quarterly* 119(2): 269–299.

Siqueira,-Gay, J., Sonter, L.J., and Sánchez, L.E., 2020. Exploring potential impacts of mining on forest loss and fragmentation within a biodiverse region of Brazil's northeastern Amazon. *Resources Policy*, Elsevier, 67( C). https://doi.org/10.1016/j.resourpol.2020.101662

Villar, D., and Schaeffer, D., 2019. Disarmament is the new war, Gold is the new opium, and Ecohealth is the historic victim. *Environmental Health Insights* 13: 1178630219862241.

World Bank and International Council on Mining and Metals, 2017. The growing roles of minerals and metals for a low carbon future. World Bank Report, Washington, DC, pp. 1–112.

CHAPTER 4

# Food as Commodity – 'Super' Food Insecurity of Indigenous Peoples

## *Analysis from Asia, Africa, and Latin America*

**Geographical Location**: India, South Asia; Bolivia, Latin America; Indonesia, Southeast Asia; and Uganda; East Africa

**Indigenous Peoples**: Quechua indigenous communities, Batwa tribe, Orang Rimba tribe, and Ladakhi communities

### 4.1 INTRODUCTION: FARMS TO FINGERS

> People are fed by the food industry, which pays no attention to health, and are treated by the health industry, which pays no attention to food.

The above quote by an environmental activist, novelist, poet, cultural critic, and farmer, Wendell Berry, reflects the way food and health are managed in modern societies.


DOI: 10.1201/9781003153061-4

Food is the fundamental right of all beings. Indigenous food systems have evolved over millennia to ensure ecological sustainability, but in modern society, it has been marginalized in science and policy. Science is projected as not a belief system, but one that builds and organizes social and natural world knowledge in the form of explanations and evidence. Indigenous knowledge (IK) is also tested in a rigorous way over several generations often shared and preserved through oral tradition. There has been increasing demand to integrate IK into mainstream society in food production. However, it is impossible to extract IK without practicing the 'way of living' of indigenous communities. Indigenous food systems safeguard ecological, social, health, and overall well-being but continue to be ignored by Western science and mainstream formal policies. Moreover, commercial agricultural and food production research often works on biased narratives that indigenous food systems block economic growth and sustainable development. The indigenous food system is a term that broadly encompasses: first, the origin – of forestry, small family farms both crops and livestock, fisheries, aquaculture, wildlife nature, and second the activities – the way food and nutrition are processed, preserved, distributed, the culture and spiritual values, sustainable production or procurement, and equitable consumption along with sustainable reuse/recycle of the waste as organic manure back to the soil.

Vijayan et al. (2022) explain, "Academic research on indigenous knowledge benefits Indigenous Peoples only if it contributes to concrete interventions that support indigenous communities in conserving relationship with local environments. [...] For example, seeds of local crop varieties in gene banks can play an important role in the transfer of indigenous food systems knowledge between regions and across generations. At the same time, ex-situ methods need to be developed carefully together with communities as they also create novel risks of exploitative bioprospecting and biopiracy." Further, they elaborate on seven entry points for engaging indigenous food systems knowledge that includes (1) documentation of the oral food systems; (2) loss and resilience both due to exogenous practices such as global food commodity markets as well as forced displacements; (3) strategies for conservation; (4) negotiating knowledge diversity; (5) representation in practices and policy; (6) recognizing co-evoltion of ecosystems and knowledge systems, and (7) indigenous self-determination. These seven entry points are basic minimum requirements, which puts an emphasis on the value of the indigenous food system. However, like several research studies (see Kuhnlein and Chotiboriboon, 2022), Vijayan et al.'s paper (2022) fails to elaborate on how science fails to recognize IK as evidence-based knowledge, or rather why only information when 'scientifically tested and proven' is regarded as science as compared to the indigenous food systems knowledge that has been evolved with 'practice and adaptation over several generations,' mostly orally, is not acknowledged.

### 4.1.1 Acknowledging Indigenous Food Systems Knowledge: A Conceptual Framework

Indigenous food systems knowledge has been historically ignored by science that benefited corporates in promoting mechanized agriculture production. Moreover, the myth that indigenous food systems cannot benefit global 'food insecurity' has never been proven by mainstream science. The linear value chain-oriented food system promoted by corporates with the aid of mainstream science differs from the intrinsic values of indigenous peoples' food system that is linked to culture and spiritual knowledge of rich biodiversity food, food sovereignty, and healthy well-being. The 2021 United Nations Food System Summit has been without controversies, particularly exposing how the marginalization of indigenous food systems is sidelined by corporate interest (Lakhani, 2021).

The importance of understanding the food system is in a way how we understand the term food. Food is a resource that would hold diverse meanings and associations for diverse sets of individuals, communities, regions, and nations. Food is a powerful driver of human agency as it relates to hunger, our need for survival. The notion of food by indigenous peoples almost across the globe differs by identifying food as 'commons' from the mainstream society's outlook on food as a 'commodity.' There are various food-as-a-commons discourses but for the purpose of this chapter, it will be about cooperation, sustainability, sharing, equity, recognition, self-governance, and not-for-profit and not commercialized. Though the food has been commercialized for many generations, in recent decades the intensity of commodification has extended to foods that were otherwise 'commonly used and protected by societies,' including indigenous peoples. In recent decades, the dominant economic discourses have been holding strong narratives that market force is the only way to regulate food systems to eradicate global hunger and malnutrition, which has been proved incorrect (Vivero Pol, 2014). However, purely treating food as a private commodity has resulted in denying millions of people – including indigenous peoples who conserve about 80% of the world's biodiversity and are guardians of food ecosystems.

indigenous peoples' concept of commons is broad; it ranges from spiritual connection to collective sharing of resources and nature conservation of not just the 'food' but entire ecosystems that support sustenance of food, for example, from earthworms to pollinators, protecting rivers and forests, to soil and air, thereby making a 'food-as-a-commons' complex web to establish the economic value of food unlike the state-protected to market-regulated agri-food sector. The market-regulated agri-food isolates food from the concept of health, well-being, and nutritious diet, unlike the food-as-a-commons. Vivero Pol argues (2014) that "markets, which are governed by self-interest, will not provide sufficient public goods, such as public health, good nutrition or the eradication of hunger, with large-scale non-monetary benefits for human beings." The market-based multinationals that are penetrating the low-income countries in Asia, Africa, and Latin America are governing food based on the economic value that offers food to those who can afford it while at the source they compete with small family farms that otherwise would be self-sufficient.

In this chapter, I explore two aspects of indigenous people's perspective about their traditional knowledge of food systems and how the marketing trend of 'superfood' by corporates is denying the right to food of indigenous peoples I but also devoid them of acknowledging their IK. The first aspect is whether commercial businesses acknowledge the traditional IK; IK here refers to not just about the technique of processing and producing a product, but the way the ecosystem is being managed sustainably. The second, impact of large-scale corporate food production and urban demand for superfoods push otherwise traditional local food away from indigenous communities.

## 4.2 THE STUDY AREA AND DATA COLLECTION METHOD

The case study areas are in India and Indonesia in Asia, Uganda in Africa, and Bolivia in Latin America which are inhabited by indigenous peoples in indigenous territories in diverse landscapes (see Map 4.1). The details of the four study areas are elaborated below. One of the common threads in selecting the study area is that the communities are practicing IK right from food production or foraging to preservation and consumption.

The primary data collection was qualitative and collected between 2004 and 2018 from India, Bolivia, Indonesia, and Uganda. The fieldwork was conducted by me with the support of community-based organizations. Each of the country cases had a similar set of questionnaires, which were qualitative with a mix of open- and closed-ended questions. About 25–30

MAP 4.1 India, Uganda, Indonesia, and Bolivia.

households were interviewed from each study village area including both men and women and youths. The semi-structured interviews were conducted in the local indigenous language or dialect with the help of an interpreter. In addition, secondary data such as gray literature, popular science articles, and local media news were referred to for supplementary information. Each case study presented below is structured with information about the principal food as a 'common' that has become a 'commodity.'

## 4.3 CASE STUDIES

### 4.3.1 Ladakhis, India

Ladakh, a northern region of India, has a strong military base and also is a sparsely populated region and is the highest plateau, about 9,800 ft or 3,000 m in the country. The case study area, Saspol, is based in Leh district. About 70% of the population in the study village were Buddhists (mainly Tibetan Buddhists) at the time of fieldwork conducted by me in 2017–2018. The Ladakhis, as the local communities, have practiced small-scale family farms for generations, without intensive use of chemical fertilizer and mechanization; most of the food produced has been for local consumption.

One of the typical breakfasts is *ngampe* (or *tsampa* in Tibetan) or roasted barley flour, which is the staple food of the region. Barley grinding has been traditionally done using the traditional stone-grinding watermill. However, the mountain villagers of the study area lost many traditional mills due to flash floods (cloudbursts) in 2010 and have remained ineffective due to loss of water flow. The watermill, known as *rantak* or *churak*, works on the rapid flow of water – often, a snow-fed mountain stream – that powers the mill built next to the stream. After the flash flood, the village continues to have many functional indigenous eco-friendly mills during summer months when the river/stream is not frozen. The villagers, over the decade, have moved farming from barley to wheat, for various reasons. One of the elderly farmers mentioned that the younger generations who often go to work for the military as laborers have developed the taste of wheat-based food and prefer wheat over barley to become assimilated with the mainstream community. The other reason, according to the respondents, links the introduction of wheat in the region to the establishment of military camps that have been distributing the grain at a free or subsidized rate for over two decades. Nevertheless, for the elderly population in the village, wheat is an alien crop, and they continue to farm barley. On the contrary, the appreciation for *chang*, a barley-based indigenous wine, has increased due to tourism in Ladakh. The traditional method of making *chang* is compromised due to high market value, which is benefited by the middle agent who buys it from local producers and sells it in Leh, the capital city.

According to the villagers in the study area, many families are cultivating barley only for market sale, rather than consumption, due to the increase in market value and reduced local consumption. "My family has been eating *tsampa* (Barley flour) that is organically produced on our small family farm and pounded in our stone-grinding water mill. Everything is sustainable. There is zero waste. Our children living in cities have lost the value of indigenous food culture, and only now with the global fad about millet as a superfood, our children are reflecting how rich indigenous food culture we Ladhakis have – from our farm to the fingers."

Both men and women of the household are equal partners in family farming. The unique collective culture of the community ensures a barter system of labor. It is typical for the communities – neighbors – to extend support during the farming season on each other's farms. This barter system of labor is the core of traditional farming for the communities in exchanging knowledge, seeds, and indigenous tools and techniques. Typically, the food produced is for self-consumption and is often stored for the winter season.

### 4.3.2 Orang Rimba, Indonesia

Jambi province in the central part of Sumatra Island of Indonesia has one of the remaining hunter-gatherers indigenous groups known as Orang Rimba (which means people of the forest). They were traditionally nomadic and now largely live in and around forested national parks. Some of them have been displaced due to the government's development policy such as road construction and extractive mining in forest areas including palm oil plantations. To some extent, the Orang Rimba communities continue to have clan-based structures, often the families are matrilocal – where husbands reside with the wife's parents. Each clan is headed by a chief and the chief has several associate leaders who have specific tasks within the clan.

Forests, for the Orang Rimba communities, have several significant contributions to the communities' life cycle from birth to burial and from farming to religious rituals – good and evil. Each clan maintains its own forest area with mutual understanding between other clans and they respect it. This traditional law, known as *adat*, has been strongly adhered especially when other clan members want to access and use the forest, they take prior permission. This case study was conducted around *Bukit Duabelas* (which means The Twelve Hills) in the year 2004 when the National Park regulations were introduced creating park zones that denied Orang Rimba access to the forest areas. This tropical rainforest has rare exotic trees that are valuable for their timber and known for some endangered flora and fauna.

During the fieldwork period, the deforestation rate of the forest was high due to timber logging, and the cleared area was used by the company and the state for oil palm plantation expansion. The elders interviewed explained that lack of tenure rights and lack of recognition of their *adat* law by the state meant corporates such as oil palm companies as well as forest concession have converted forests into oil palm plantations. The *adat* law is recognized by the state at the time of writing this chapter. The focus of this case study on deforestation and lack of tenure rights and its impact on Orang Rimba remains valid. The prime impact on Orang Rimba was the loss of their traditional culture of hunting and gathering due to the Trans-Sumatra highway resettlement, lack of tenure rights, and influence of the forest industry promoting large-scale oil palm plantations.

Traditionally, Orang Rimba has largely been dependent on forest wildmeat, which has been a major source of their nutrient intake. However, due to resettlement, their access to forest wild meat has been reduced to negligible. One of the interviewees expressed fear, "We used to hunt small non-endangered species for food consumption and have maintained forest wildlife conservation. With the change in land use, our livelihood sustenance from the honey collection and other non-timber forest product gathering such as roots and tubers has declined. We can hardly find a wild boar and when we do we share within our clan. To survive, we must buy staple food, which used to be forest commons earlier."

The communities, according to the women's focus group discussion, were being pushed to a sedentary lifestyle and to assimilate the mainstream, including religion. Many are forgetting the indigenous food system that included food taboos and customs with an important function, to respect the seasons – the breeding period for both fishing and hunting. Food taboos were more to avoid contamination and over-exploitation of commons than for any religious reasons. One elderly woman expressed, "Earlier our girls were mostly indoors and rarely interacted with outsiders, but things are changing. Due to changes in our food system, we are forced to do family farming. Cassava used to be our main staple, but now due to lack of access to the forest we are unable to forage cassava, so we are adapting to eating rice, which we have to buy it." The otherwise wild food that we consumed is now hunted by settlers who sell it for profit. The Orang Rimba elder Chief said, "Our last three generations have suffered from loss of indigenous knowledge of food system because when you take away our forests you take away our knowledge of food as medicine (herbs, shrubs), food as nutrients (proteins and vitamins via wild animal meat), and food as our taste (mineral rich organic wild roots and tuber instead of buying white rice)."

### 4.3.3 Batwas, Uganda

Batwa people, like Orang Rimba, are traditional hunter-gatherers who used to be nomads moving across the forest areas bordering Uganda, Rwanda, Burundi, and the Democratic Republic of the Congo. This case study is of Batwas living in and around Bwindi Impenetrable National Park, Uganda. The ethnographic research was conducted by me in 2014. Bwindi forest is located in southwestern Uganda and on the east of the Albertine rift valley. The Batwa are considered to be the first inhabitants of the forest in Central Africa and their population inside Bwindi is approximately 1500.

Within Bwindi, they were evicted and restricted from accessing the forest. According to a Batwa elder, he and his family received no compensation or any resettlement. They live in and around the National Park and some men are hired as porters. At the time of fieldwork, there were discussions that because gorilla tourism makes huge profits it should engage Batwas for their IK and inherent ability to track gorillas. However, the argument given was forest conservation and tourism departments need youths with formal education.

One of the village or hamlet Batwa leaders explained, "When foreigners come to Bwindi, the forest rangers hire us as porters to carry heavy luggage to the campsite and during sightseeing. However, what foreigners do not know is that these forest rangers who are not Batwas know nothing about the forests. The rangers ask us about different types of flora and fauna and their uses, which they translate to foreigners in English as if it is their own knowledge from education. It is our indigenous knowledge they exploit." The Uganda Wildlife Authority officer whom I met during fieldwork mentioned, "It is a complex situation given that Batwas do not have any land rights and education degree, they are given the job that suits them such as porter. Moreover, if we hire them, they will resort back to hunting and gathering in the forest area, which will not serve the purpose of helping them to be sedentary farmers." The Batwa youths in the focus group discussion raised their concern about their lack of knowledge of how much they should earn, and they feared any demand for workplace safety and health insurance will mean loss of job.

The women in the group discussion expressed that they are sexually harassed by non-Batwa men and by outsiders for sex, often forced to do in exchange for food and/or money. Being displaced from the forest and deprived of forest food, Batwa men in the study area have fallen trap of middle agents who sell them alcohol in plastic pouches. There were plastic pouches all over the village and on the small farms, which has not only created environmental challenges but taken a toll on many lives. A Batwa woman mentioned, "Earlier our elders use to live long, but now in this village, many are unable to adapt to the lack of wild edible food. They suffer from health issues due to lack of nutrition and are lured to alcohol often provided in exchange for labor work. It (alcohol) helps to forget misery, but it is not food. Our forest is taken away from us, and now we find our indigenous knowledge useless, and we have to buy food (as a commodity instead of collective commons)."

During the focus group discussion, the youths mentioned that there is a huge demand for bushmeat outside the park. Often the small animals are illegally hunted and sold, and the blame is put on the Batwa communities given that they have a tradition of eating wild bushmeat. "We have nothing left after the displacement, but we continue to be on the receiving end of losing our food as commons and being blamed for others' choice of making our food superfood. This worries us to continue to live in the vicinity," says one of the Batwa youth who works with civil society that aims to empower his community.

### 4.3.4 Quechua, Bolivia

In 2013, I conducted fieldwork in Bolivia's Cochabamba. Quechua are indigenous peoples living in the Andes Mountain region of Bolivia, Peru, Ecuador, and Colombia. Historically, Quechua have faced displacement from their collective land, and attempts are being made to assimilate them into mainstream society – influencing their language, culture, and way of living. The two mountain villages in Cochabamba had unusual frost that left many livestock mainly alpaca herds and llamas dead. Harsh terrains and low temperatures destroyed food sources and loss of crops.

The study area is in the highland and maintains one of the largest varieties of potatoes. Indigenous women were the important guardians in protecting and conserving biodiversity and seed preservation. From observation, it might look like a gendered division of labor in food production – men taking care of the farm while women sold the products in the market; however, what I found was a bit more equitable. One of the biggest challenges for them is their traditional food *kaniwa* which is one of the hardy crops along with quinoa. These traditional crops along with potatoes have been a staple food in the Andean diet. Due to climatic conditions and IK, Quechua people have managed to produce and maintain the diversity of potatoes using minimum water. These potatoes, according to a Quechua woman, have good nutrient value compared to factory-produced potatoes. She said, "Our potatoes may not look as white and beautiful, but it has a high amount of vitamin and mineral that is essential to survive in this weather. Even potato scientists cannot replicate the nutrients when they take our potatoes to the lab, why? Because it is not just manufacturing potatoes, what we do is take care of Pachamama (mother earth); farming for us is about connecting the soil with the soul."

The diversity of potatoes and traditional ecological knowledge of potatoes such as how to preserve and grow in harsh climates and how to care to make them resistant to pests and diseases are passed on not just from across generations but also across communities. Conservation of the indigenous food system encompasses all elements of the earth, which the Quechua communities consider an important part of their knowledge. "You cannot pick up one aspect of our food system and ask how does this potato resist frost or why *kaniwa* is our kitchen crop? Yet, when we sell our potatoes, we do not get higher value for preserving our knowledge," says a Quechua women leader who is the top seller of potatoes in the study area. On the contrary, the demand for indigenous varieties of potatoes, quinoa, and *kaniwa* yields a high value for being a superfood, which pushes the local communities from eating it as a staple food and rather selling it in the market for money.

## 4.4 DISCUSSION

The diversity of the indigenous food system across the landscapes and countries is evident from the above-mentioned four case studies. Table 4.1 highlights the significance of food culture for indigenous peoples, examining the way food has transformed from commons to commodity and the impact of the commodification of the food to market as a 'superfood.' Using the analytical framework, I discuss the greater recognition of indigenous rights and governance for stewardship of traditional land and food systems. The case studies provide evidence as to how these principles should be achieved including the acknowledgment of indigenous food from cultural to ecological importance.

### 4.4.1 Freedom to Farm

Farming is evolving, even for the remotely located indigenous communities. Local food from farms, though being promoted, in the last century industrial food system has curtailed freedom to farm through the privatization of seeds, fragmenting the idea of collective land, excessive legislation on market and price, adding patents by denying rights to all, and used publicity to promote indigenous food as 'superfood.' As I traverse between the study areas from Bolivia to India (see Table 4.1), it is obvious that indigenous communities have been practicing farming that is common in a way that either the land is collectively managed as in Quechua or in the case of India the farming was organized collectively to support each other in producing food and nutrition.

TABLE 4.1 Comparative Overview of Food Production and Consumption among IPs in India, Uganda, Indonesia, and Bolivia

| Indigenous Peoples/Food Culture | Ladakhi, India | Orang Rimba, Indonesia | Batwas, Uganda | Quechua, Bolivia |
|---|---|---|---|---|
| **Farming** | Collective farming | Forest farm, hunting, and fishing | Forest forage/ hunting-gathering | Collective farmland |
| **Traditional (heirloom) produce** | Barley, heirloom root vegetables radish, carrots, turnip, and spinach | Sago, cassava, *gadung* (a tuber), and fruits and honey | Not applicable | Quinoa, heirloom potatoes, and *kaniwa* |
| **Preference for heirloom crop production** | 1. Historical link with the introduction of the crops, 2. Adaptability to the harsh prevailing environment, and 3. Storage of the food for longer (winter) period | | | |
| **Knowledge for conservation** | The seeds are locally conserved and used | The seasonal respect for hunting and fishing | Co-existence with wildlife and nature | Soil fertility conservation is critical due to tough terrain |
| **Development dilemma** | Food assimilation with non-local culture | Modern farming vs traditional ecological knowledge | Decline in forest and land access | Adopting new climate-resistant food varieties thereby losing traditional food |
| **Sustainability – coping strategy** | Community barter system | Reviving traditional farming | Adapting to mainstream | Migrating to city (La Paz) |
| **Vulnerability** | Cultural assimilation | Monoculture plantation | Eviction from forest/relocation | Lack of land tenure |

To understand and acknowledge the IK, the case studies highlight that IK is not about patents, but sharing the knowledge so others can benefit, it is not about maximizing profit, but ensuring self-sufficiency for everyone, it is not about consumers eating food, but about the culture of appreciating through festivals associated with human's life cycle from birth to death. The IK by linking our life with the farm creates a bond, while this bond in the industrial food system not only lacks but by distancing consumers from the food creates fear about indigenous food systems as unsecured.

### 4.4.2 Transition Toward 'Just Commons' Food System

A just transition again toward food as common is a good idea, but whether it is doable demands a different answer. The current food system has created an environment that makes us rely on corporate monopoly. For indigenous food systems to get recognized and implemented widely, a new form of food governance becomes critical such as one where the main emphasis is on the right to food and well-being. The case studies indicate that agricultural innovation for food security should rather be scaled down to promote a common and collective food system based on IK.

For a 'just common' transition to happen for the food system, it is important to identify the different types of social, legal, political, or spiritual vulnerabilities that indigenous communities are facing. For almost all case studies, a common factor that pushed away the communities from the foods that otherwise were common to them was linked to assimilation with the mainstream societies. The assimilation creates a disparity in the identity of indigenous culture, and the relocation or displacement adds extra pressure on the communities' sustainable food systems and food habits. The finding from the four comparative case studies corroborates the study by Warne and Wescott (2019) wherein they found that forced relocation of American Indians from ancestral lands limited their access to traditional food, which made them dependent on food subsidies, which contributed to poverty, alcohol, and substance abuse. The transition toward a 'just commons' food system could be realized when the indigenous peoples are understood as a part of the environment where they belong such as Orang Rimba, Ladakhis, Quechua, or Batwa are inseparable from their forest food system because that is how they identify themselves.

## 4.5 CONCLUSIONS AND RECOMMENDATIONS

Indigenous peoples' food systems can be realized with the recognition and through maintaining cultural, social, spiritual, and forest and land rights. In conclusion, I would like to highlight that the contribution of indigenous peoples' food systems should not be taken for granted, rather it should be respected with free, prior, and informed consent (FPIC). Their knowledge when combined with Western science can only strengthen forest food policy and land-use decision-making. Food is not just a consumable product for indigenous peoples, but it is a pride – of how they produce, and therefore stewardship is critical.

A few key policy and practice recommendations would be to assess the commodification of food and monopolization of the food system that is promoted by Western science for food security. The indigenous peoples' food system should not be considered an option, but rather a mainstream solution that opens up the sharing of knowledge and exchange of seeds to traditional technologies. The commons could be reviewed by addressing the management, access, and forest land tenure rights of indigenous peoples. Another recommendation would be to strategically shift the paradigm of colonialism and instead promote food sovereignty – recognizing and amplifying the voices of those who are food caretakers and knowledge keepers across the globe.

## REFERENCES

Kuhnlein, H.V., and Chotiboriboon, S., 2022. Why and how to strengthen Indigenous peoples' food systems with examples from two unique Indigenous communities. *Frontiers in Sustainable Food Systems* 6. https://www.nature.com/articles/s43247-022-00543-1

Lakhani, N., 2021. 'Corporate colonization': Small producers boycott UN food summit. *The Guardian*, accessed online on 17th May 2023 via https://www.theguardian.com/environment/2021/sep/23/small-producers-boycott-un-food-summit-corporate-interests

Vijayan, D., Ludwig, D., Rybak, C., Kaechele, H., Hoffmann, H, Schönfeldt, H.C., Mbwana, H.A., Rivero, C.V., and Löhr, K., 2022. Indigenous knowledge in food system transformations. *Communications Earth and Environment* 3(213). https://doi.org/10.1038/s43247-022-00543-1

Vivero Pol, J.L., 2014. The food commons transition - Collective actions for food security. The Broker, accessed online on 15th May 2023 via https://www.thebrokeronline.eu/the-food-commons-transition/

Warne, D., and Wescott, S., 2019. Social determinants of American Indian nutritional health. *Current Development Nutrition* 3(2): 12–18.

CHAPTER 5

# Pastoralists, Nomadic Movements, and Identity in Tropical Grasslands

**Geographical Location**: Uganda, East Africa; and India, South Asia

**Indigenous Pastoral Tribes**: Bahima tribe and Maldhari tribe

## 5.1 INTRODUCTION: PASTORAL NOMADIC TRIBES

Pastoralism is a broad concept that commonly refers to the practice of taking care of (domestic herbivores) livestock as a means of either primary or secondary subsistence. Normally, pastoral societies manage ruminant animals that have natural herding habits. Pastoral nomads have been practiced for generations, particularly in semi-arid dryland ecosystems. Most traditional pastoral nomads continue to practice in their wild ancestral forms in tropical savanna and in temperate grasslands. According to Dyson-Hudson and Dyson-Hudson (1980:15), pastoral nomads are those people who are principally dependent for their subsistence on livestock, and for whom spatial mobility is regularly employed as a survival strategy. Yet, it is challenging to categorize pastoralists as a homogenous community because the spatial mobility among them differs in diverse landscapes and for the type of livestock.

From as early as the 1950s, scholars have tried to categorize the pastoral nomad based on how much time they move with their animals for green forage (Awad, 1959; Dyson-Hudson and Dyson-Hudson, 1980).


DOI: 10.1201/9781003153061-5

The factors and causes associated with such categorization have over time become blurred between each category. Based on the literature review, a few identified categories are as follows (Bacon, 1954; Clarke, 1959; Falah, 1990; Little, 2016): transhumance pastoralism is a category for pastoralists with fixed seasonal movement of livestock with change in altitude, often vertical movement in the mountains. Semi-nomadic are those who often plant a few crops at their base camp before moving out on the seasonal migration, while semi-sedentary referred to those who live in permanent villages for a part of the year, where they plant crops, and move out in tents only during one season of the year. True nomad pastoralists are regarded as those who move from one place to another, often horizontally and across geographical boundaries. It makes it impossible to find any features of culture or social organization that could cut across all types of nomadic categories (Spooner, 1971).

Almost half of the 1.4 billion global cattle population is in Asia and Africa despite the proximity of wildlife to domestic stock (Little, 2016). The 'cattle complex' is the concept used to describe the system of values that governed native cattle ownership in a large part of East Africa (Herskovits, 1926). The concept highlights that cattle-keeping is not just about economic importance but brings cultural meaning to the life of pastoralists, mainly the prestige system related to cattle ownership has been restricted to men. Marriage required the transfer of cattle in the form of wealth either to the woman's family (Kenya) or to a man's family (India). It is obvious that despite the diversity and geographical distance among pastoral nomadic communities across the continent, one factor that binds them apart from their characteristic of happening to be brave, independent, and freely moving with their herd is their traditional knowledge that is uniquely intertwined with mobility and livestock management.

A few recent studies on indigenous knowledge of nomadic pastoralists attempted to show that the Somali nomadic pastoralists in Kenya did not translate the knowledge into safer health practices, for example, Muga et al. (2021) highlight the Rift Valley Fever, a zoonotic disease that not only affects livestock but also people and wildlife, which could cause heavy economic and public health burdens, while a study by Barnes (2020) in Mongolia shares similar findings that most households had knowledge but a few had recognized the risk of reverse zoonoses or human-to-animal disease transmission. Though such inventory research studies are useful, it fails to delve into the holistic understanding of eroding indigenous knowledge due to land-use changes nor applauds the household livestock holdings by African and Asian pastoralists who have faced limited exposure to diseases.

Globally, there is a tendency to discourage pastoral nomads because of some of the prevalent myths that continue to float around such as listed by Farvar (2003:31–39) that pastoralism is an archaic form of production not adapted to modernity, mobility is inherently bad, all conflicts are caused by pastoralists, rangelands are degraded and have become 'wasteland' because of over-stocking and overpopulation, pastoral production has very low productivity, pastoralists have to settle down in order to benefit from modern services, pastoralism destroys biodiversity and leads to desertification, pastoralism does not take care of the land because of the 'Tragedy of the Commons,' pastoralists are poor, in ill health, and food insecure, and pastoralists overgraze their land.[1] Across the globe, from Artic pastoral ecosystems to tropical grasslands, these myths continue to favor the policy decision-makers to divert the pastoral land for extractive industries or to corporate companies that put multiple competing commercial land-use pressures on pastoral farming (Forhinger et al., 2021; Sternberg and Chatty, 2016). Authorities, when conventional policies and planning fail to settle pastoralists, continue their efforts for settlement with indirect methods, such as the construction of roads, railways, or canals that hinder the nomad from their traditional practice of moving with their herds and practicing indigenous knowledge of several generations (Awad, 1959; Nori, 2022).

Various countries have formulated policies encouraging settlement with an argument for the economic and environmental development of 'wasteland,' the challenge to implement diverse regional policy frameworks to meet the diversity of herders, and for social 'assimilation' with the mainstream population. Most policies, laws, economic incentives, and private company investments do more harm to pastoral nomads than to harness their potential. The marginal recognition of pastoralists in national policies is evident in tropical grassland countries. Such exclusion has led to distinct evolutions of national and regional policy frameworks that further hinder the pastoralist's traditional cross-border moves and knowledge exchanges for fear of losing the indigenous knowledge.

The aim of this chapter is to examine indigenous knowledge that is intertwined with pastoral identity and the herding movement or migration and its impact on the forest and land-use policy decisions in Uganda and India. Including this introduction, the chapter is divided into four sections. In section two, I discuss the study area and primary data collection methods. In the third section, the empirical evidence and the findings of two case studies – from South Asia India's Maldhari and East Africa Uganda's Bahima pastoralists – are presented. The fourth and final section discusses using the evidence from the two case studies the role of the 'identity' of pastoral nomads and the 'herding movement,' in forest and land-use change decisions, and concludes with key policy recommendations and research priorities for the future.

## 5.2 THE STUDY AREA AND DATA COLLECTION METHODS

I collected ethnographic primary data during fieldwork in Banni grasslands of Kutch district, Gujarat, Western India, and Lake Mburo National Park, Mbarara, in South-Western Uganda (Map 5.1). The fieldwork involved individual interviews with pastoralists based on questionnaires as well as four focus group discussions between 2013 and 2016. Of the four focus group discussions, two were exclusively for women and youths to include diversity. Twenty-four pastoralist families in Banni and 15 cattle keeper households – who could be classified as semi-nomadic and semi-sedentary – were randomly selected through a combination of hamlets/clusters

MAP 5.1 Map of Gujarat in India (Kutch, Banni grassland).

and stratified sampling procedures considering geographic distribution and demographic as well as the economic situation of the household.

The semi-structured questionnaire consisted of both open and closed questions, which were translated into the Sindhi language for the Maldhari tribe and Runyankole language for the Bahima tribe. The semi-structured questionnaires covered topics including how indigenous knowledge related to breeding strategies, herding movement, and management practices of both grassland and the biodiversity and breed of herds is understood in the community, and how the knowledge is preserved and transferred to others. To engage the pastoralists about their perceptions of policy reforms affecting the forest and land-use change, the focus group discussions were held on different topics such as human–livestock–wildlife co-existence, the effect of herding movement due to the status of the national park and reserved grassland, and changes in traditional collective rights to land-use and nature conservation.

## 5.3 CASE STUDIES: MALDHARIS, INDIA AND BAHIMA, UGANDA

### 5.3.1 Maldharis, India

Originally, nomadic pastoralists, the Maldharis is an occupational term for Muslim nomadic pastoralists. *Mal* means wealth or livestock and *dhari* denotes owner or keeper, who are traditional dwellers of Banni grasslands, Kutch region of Gujarat, Western arid India. Different sects of Muslim communities comprise almost 90% of Banni's population, and Meghwal Hindus are distributed in some 50 hamlets of 19 village Panchayat (village-level democratically elected bodies). The intermingling of both communities provides a symbiotic relationship and continues with the age-old tradition of bartering – often, exchanging milk, leather, cultures of playing musical instruments and songs, and exquisite embroidery skills known as Kutch embroidery by women. A diverse sub-sect of Muslims including Halaypotra, Hingora, Jat, Node, and Mutwa tribes have settled in Banni from as far as Afghanistan and the Sindh region of Pakistan, primarily looking for good fodder for their livestock.

Banni Grasslands is Asia's largest grassland and is regarded as the finest grassland in India. It is an arid grassland ecosystem and during the princely rule of Kutch rulers, it was allocated to pastoral communities to graze and breed their livestock (refer Map 5.1). In 1955, it was legally declared a protected forest under the status reserve forest in India, under the Indian Forest Act of 1927. Over the years, Maldharis have become semi-nomadic herders who travel with their livestock including Banni buffalo breed, sheep, goats, cows, and camel – in particular, the Kharai camels, also known as swimming camels of the mangroves living in Banni area. Both Banni buffalo and Kharai camels are regarded as separate breeds, which makes this grassland and Maldharis gain special recognition globally. To ensure their grassland is recognized, they collectively formed Banni Pashu Uchherak Maldhari Sangathan to conserve their breed and maintain the grasslands of Banni. This collective action is even more important given that the state fails to recognize pastoralism as an official 'identity' category.

### Grassland vs Fortress Conservation

The concept 'fortress conservation' is used to define an approach that biodiversity protection can be only achieved by creating protected areas that (1) exclude traditional local communities and their livestock, (2) promote tourism or safari hunting, and (3) deny recognition of indigenous knowledge instead of Western fortress model of conservation. One of the village elders, who is highly esteemed by the community, with immense indigenous knowledge of animal husbandry, the late Salim *bhai*, explains, "Even before the forest officials arrived here to declare as Conservation Reserve, our Chhari Dandh (wetland for migratory birds in Banni), we have been protectors of these birds and know exactly what kind of environment they favor.[2] The officials introduced the non-indigenous weed, *Ganda Baval* (*Prosophis juliflora*), our Banni began to degrade. When we complained that our cattle are suffering from eating the pods of *Ganda Baval*, the officials called us ignorant and recommended we stop nomadic pastoralism, without realizing our Banni buffalos are world famous due to its herding movement and grassland." In 2010, the Banni breed was formally recognized as the 11th breed, the first one after India's independence, which according to the respondents was an important step.

The focus group discussions with Maldhari men revealed that instead of preventing the salinity expansion to growing near the vicinity, *P. juliflora* is like a weed denying traditional grass, increasing the toxicity of the soil and spreading rapidly all over Banni. Despite several pleas by the pastoral community to consider their indigenous knowledge for the long-term conservation of the Banni, the official did not pay heed. The federation suggested the officials for removing of *P. juliflora* and instead promote the grasses that are local and favored by their cattle such as *Drab* (*Desmostachya bipinnata*) or *Dhaman* (*Cenchrus ciliaris*). The long list of local grasses was developed as part of their indigenous knowledge – mostly oral tradition and was shared with the local authorities. The local authorities, however, considered that Maldharis has been at fault for overgrazing causing the reduction in the grass quantity and diversity because of the herding movement. "Our livestock are getting sick of forest officials' ignorance toward our indigenous knowledge. They bring scientists from the cities who have western degrees, but lack local knowledge, which brings disasters to us because they do not listen to our elders and our knowledge," said the leader of the Banni Breeder's Association.

"We must frequently patrol the wetland that the official declared as Conservation Reserve. The authorities have to spend money on putting up barbed wire fences to declare the protected area boundary and any trespassing would result in fines and punishment for the entire community. What good it has done to migratory birds in Chari Dhand other than bringing more urban tourists who understand nothing about the local culture? The bird watchers think the fence will be good for the bird's protection, but they do not come to see the impact of the barbed wire. When big migratory birds like Houbara bustard try to take flight, they need a big runway before taking off. They do not see the fence and in the process often their wings get tangled in the barbed wire fence, hurting them. We go to take care of these birds!", according to Halim *bhai*, who takes a special interest in birds and often acts as a local guide to spot migratory birds' arrival and nesting area. The fortress conservation meant that the local guides, like Halim *bhai*, are excluded from the state-promoted tourism in the area – thereby declining the oral traditions of storytelling (Daastan) and the significance of birds in Maldhari's songs of various forms – showcasing the amalgamation of Maldhari's diverse cultural background.

### Herding Movement, Climate Change, and Banni Buffalo

The Banni buffalo is important livestock for being recognized globally as having excellent germplasm and maintaining a unique breed. On average, Banni buffalo cover about 10–20 km distance mainly grazing at night without any human assistance. Maldharis have a unique system of herding – each family's lead buffalo has a bell on its neck, a unique bell sound that helps the herd identify whom to follow and return to the same household in the morning. Only during summer, the Maldharis walk buffalos to different locations, while other small livestock such as sheep, goats, or camels is herded regularly. The Maldharis, including the youths, can navigate the wide arid area of the Banni without any technological navigation assistance. This immense sense of geographical orientation is based on their indigenous knowledge of the landscapes – for non-locals, such distinctions will be invisible.

According to the respondents, Banni buffalos are adapted to frequent droughts and are well adapted to the limited but quality grasses that the grassland offers, including the herding movement. Stall feeding of the same grass will not yield the same result, because the uniqueness of the breed is based on their herding movement.

Climate change is apparent, according to the Association members, but it is the fortress conservation and lack of formal collective forest tenure rights that are bigger challenges. The herding movement allows indigenous knowledge transfer among diverse Maldhari communities – about water bodies, green forage, zoonotic diseases and ways to tackle them, and the exchange of breeds for ensuring the genetic germplasm. The individual interviews revealed that the Forest Officials' narrative about 'overgrazing and nomadic pastoralism' was nothing but based on myths and damaged their otherwise ecologically sustainable livestock-rearing practices. The stall feeding and banning herding movement will steal away the pluralistic culture of Banni's ecosystem as well as the pastoralists' lifestyle. Instead, it will fragment the community and steer toward a market economy rather than a bioeconomy that is socio-ecological ethics. Moreover, the focus group discussions with Maldharis highlighted those farm animals in Western societies and other parts of India are at high risk of exposure to zoonotic diseases compared to pastoral populations. The Banni buffalo and camel herding movement is part and parcel of indigenous knowledge and pastoral identity.

### Pastoralism Identity: Maldharis

The disappointment with authorities' ignorance over the traditional governance of grasslands based on Maldhari's traditional oral history has left pastoral communities in fear of losing their identity of being pastoralists. I encountered Maldharis protesting during fieldwork to claim their collective rights to the entire grassland, which is not yet recognized (as of the year 2023). The protest, however, was beneficial because the state initiated demarcating the Banni grassland area. This mapping of Banni grassland could have a dual impact – positive, it will avoid any encroachment by private companies or other stakeholders, and negative, mapping could be counterproductive by pushing Maldharis to remain inside the demarcated area and cut their movement. Also, the state offered Maldhari families a few hectares to settle as sedentary herders, which is key to their resistance to the representation and identity of pastoralism. The injustice among pastoralist communities in Banni is mainly due to the loss of land through fortress conservation-inducted exclusion from protected areas that have only led to ecological fragmentation. Maldharis are proud of their unique identity – of being pastoralists – instead of being known by their religion Hindus or Muslims (a volatile topic in contemporary India) or the diversity of sects within the communities. The identity is closely connected to the knowledge of identifying and categorizing different varieties of grasses found in the vast pastureland, diverse types of soil and predicting the weather based on ecological changes, availability of water, and importantly an ability to transfer this knowledge to the youth.

Such identity could be lost, the focus group discussions revealed, because the mainstream discourse projects pastoralists as illiterate communities who lack the skills or knowledge to lead the lifestyle of the mainstream population. "We are offered individual land tenure for farming and some piece of land in Banni for grazing, which we declined. The minute you divide Banni you are dividing the diversity of pastoralists – which is a lifeline for us – being dependent on each other. I see the politics of divide is in work to fragment us, and hopefully, via our affiliation with the Banni Breeders Association, we can remain united, which is essential for our Maldhari identity," argues Azeem *bhai*, one of the elderly pastoralists.

### 5.3.2 Bahima, Uganda

There are a number of indigenous peoples in Uganda although Ugandan Government does not specifically recognize them as indigenous peoples. Pastoralism in Uganda is seen as a challenge; therefore, the government's plans to abolish nomadic pastoralism to embrace sedentary farming for self-sustainability are being pushed. In the past, the unique significance of Uganda's rangeland had remained unrecognized, and with land parcelization and individual tenure, the pastoralists were exposed to natural calamities such as droughts and disease outbreaks resulting in livestock mortality. "Our ancestors were displaced and had internal conflicts due to competing land uses that had reduced the availability of rangeland for herding. There are also narratives that Bahima men conducted cattle raids on other pastoralists and farming communities and due to possession of cattle we were higher in rank order of social class pastoralists. Women played an important role in the family as the 'bride price' meant more cattle for her maternal family while she became her husband's property. Things are different with sedentary pastoralism," explained one of the pastoralist women, during the focus group discussion. The Bahima are famously known for the breeding of longhorn Ankole cows. Ankole longhorn cows are an indigenous cattle breed and one of the oldest that can forage – like Banni buffalos – on poor-quality vegetation and can live off a limited amount of water, which makes them withstand Uganda's semi-arid weather. The longhorns of these cows are believed to play a crucial role in blood circulation to adapt to dry and extremely hot weather conditions.

The government of Uganda has passed several laws that promote sedentary pastoralism for Bahima tribe, which is one of the causes of their current existential crisis. The crisis is from culture to ecology: language is disappearing, the variety of fruits and honey production is reduced due to the cutting of trees and bushes, loss of trees over the decades has resulted in less rainfall but the frequency of flash floods has increased causing damage to food and livestock, and the other challenge is that migration movement has been reduced, which means limited rangeland for a large number of livestock often resulting in resource crisis, particularly water for livestock, but also conflicts within the community.

**Grassland vs Protected Areas**

Lake Mburo National Park in the Kiruhura district of the Mbarara region is a national park in the western region of Uganda (refer Map 5.2). "The pastoralist community living around the Protected Areas had a history of displacement due to Uganda's land-use policies such as undermining the value of the common resource and collective governance and practices of the local institutions or extending the boundary on the rangeland by declaring it as Protected Areas," explained one of the interview respondents from academic institute working with the pastoral communities. The dialogue during the focus group discussion highlighted the significant role of the grasslands and the indigenous knowledge of the pastoralist in maintaining the soil that stores the carbon for a longer time. One of the elders mentioned, "When a western scientist sees a grassland, they call

MAP 5.2 Map of Uganda (Mbarara region in Western Uganda).

it ‘barren’ or ‘wasteland.’ We, the pastoralist, on the other hand, identify grassland as a landscape that has evolved thousands of years of extremities and become a resilient space for the survival of grasses, and some savanna trees creating a stable ecosystem.” One of the local civil society personnel mentioned that Western scholars continue to focus on a myopic vision of protected areas in Eastern Africa.

Riggio et al. (2019) literature review paper, for example, questions the effectiveness of protected areas; the only time pastoralist communities were mentioned was in a negative narrative lack of land conversion does not mean that these landscapes are unpeopled as they may contain pastoral communities with large numbers of livestock.

These narratives are only indicators, according to the Bahima pastoralists, for denying their traditional land use of the grasslands in favor of the protected areas. The interviewees disagreed that protected areas in the fortress conservation approach by excluding them are the best approach. Moreover, they expressed that 'involving them' to get their free, prior, and informed consent is also not viable if the state's agenda is to keep the pastoralist out of the rangeland or restrict access to the rangelands.

### Herding Movement and Land-Use Change

Rapid urbanization combined with the demarcation of the common land into privately owned property has restricted the herding movement by fragmenting the collective land governance. Herding movement of the Ankole cows enables them to be resilient and resistant to any zoonotic diseases, which is declining with the sedentary and limited scope for the Ankole cows to walk and graze in longer distances. Traditionally, milk production from Ankole cows was for subsistence, for consumption by the family. Increasing demand for milk has also pushed some pastoralist families interviewed to stall feeding the cows for higher milk production.

During the fieldwork, I was told by one of the focus group discussions that emerging land-use change is subtle and often in negotiation with the Bahimas. In the name of protected areas and maintaining wildlife biodiversity, the local authorities often propose new rules or guidelines encouraging Bahimas to become active stakeholders. The duality of the land-use change decision is such that the community feels they are part of the negotiation and have been consulted but in practice, the Bahimas are devoid of power and often excluded from decision-making.

### Pastoralism Identity: Bahimas

Like the Maldharis, there is a history of sedentarization of the Bahima pastoralists in Uganda. During the fieldwork, one of the focus group discussions highlighted the identity challenge due to the introduction of settled small-scale family farms. The Bahima tribe is unfamiliar with agriculture, and from the focus group discussion, it was highlighted that the traditional cultivators are often hired for cultivating cassava or potatoes, or other vegetables. Learning a new technique such as farming demands new skills and time for which the Bahima has no indigenous knowledge. The more the community gets involved in adapting themselves to farming, the less time the younger generation learns about pastoralism. The way an increasing number of undesirable crossbreeding has happened and the younger Bahima pastoralists selling off old herds of pure Ankole cows in exchange to buy new different breeds of cows have a direct impact on their outlook as a pastoralist.

The Ankole breed of cattle is what gives Bahima pastoralists an identity and their main means of subsistence. The crossbreeding of Ankole cows is not only threatening Bahima's symbol of pastoralist identity but losing the breed is regarded as the biggest challenge as was highlighted in the focus group discussion. The fear of losing the best genetic breed of Ankole cow was echoed by women and elderly male pastoralists. The women Bahima's identity is deeply rooted in the special role they play in milk preservation. The indigenous knowledge of how to preserve the milk in wooden pots and gourds that are fumigated using special grass. The identification of these sweet-smelling grass for fumigation is the knowledge that is linked to the resources available in their landscapes. The Bahima women's knowledge to churn milk and preserve it in the most ecological and organic way could be an important learning for Western science that is heavily preservative and chemical-induced.

The sedentary lifestyle combined with education and the use of technology such as mobile phones is changing the way younger men and women from Bahima tribe continue to identify themselves. Assimilating to mainstream norms, according to the women interviewees, is worse to maintain indigenous knowledge. Women churn milk to produce ghee that is sold in the market for commercial purposes often compromising on the traditional techniques of preservation that Bahimas are known for. The cash is used for educating children in schools that fail to recognize the significance of Bahima's livelihood and indigenous knowledge thereby alienating the younger generation from their identity roots of who they are.

## 5.4 DISCUSSION

Traditional nomadic pastoralism continues to exist because of immense indigenous knowledge that has been shared from one generation to another and from one pastoralist community to another. The two cases – Maldharis and Bahimas – from two different continents show immense similarities and scope for South–South learning. The challenge is Western scientists denying the indigenous knowledge of the pastoral communities with scholars like Riggio et al. (2019) who recommend the expansion of protected areas in Eastern African countries, and I quote them:

> Where a significant portion of the region is already set aside as PAs and the remaining areas are in pastoral systems, more innovative forms of protection may be needed.

Innovative forms of protection cannot be decided by Western scholars without involving local pastoralists like Maldharis and Bahimas and their immense indigenous knowledge. Agrawal and Saberwal (2004) provide a comparative explanation of ways pastoralists manage to negotiate critical resources in India unlike countries in Africa where the representation of herder–state relations is that of steady pastoralist destitution. One of their arguments is that pastoralist groups in India are powerful vote banks and influence political decisions, which holds true for Maldhari pastoralists in this study, who are the majority in the geographical location of Kutch, Banni grasslands of Western India. Agrawal and Saberwal (2004:43) explain that Indian pastoralists in general "are better able to participate in the political process than herders in many parts of sub-Saharan or Sahelian Africa. This may partly be explained by political processes in pastoralist Africa which do not have the geographic 'reach' that the Indian system does, thereby denying many pastoralist communities much of a voice in the shaping of their own destinies." The dire situation of pastoralist communities facing land encroachment and loss of the role of traditional and indigenous local knowledge is not limited to the global south but is experienced by the Sámi reindeer herders of Sapmi in Scandinavia. Forhinger et al. (2021) studies highlight that in northern Sweden reindeer herding community's grazing grounds are functionally unavailable to reindeer due to multiple competing land-use changes, particularly large-scale mining. This is affecting pastoralists' adaptive responses due to a lack of traditional knowledge as a solution to emerging new situations such as mainstream energy and mineral infrastructure development.

One of the other striking aspects of pastoral societies and their identity is the way they view their rangelands. Maldharis and Bahimas practice semi-nomadic pastoralism and have traditionally adapted to harsh climatic conditions and institutional challenges not limited to fragmentation of pastures, loss of forage, and restricted animal movements to sustain their livestock and its unique genetic value. In the future, however, their traditional and indigenous local knowledge might be insufficient to adapt to changes that they cannot control.

## 5.5 CONCLUSION AND RECOMMENDATIONS

This study showed that the option for sustainable pastoralism is shrinking. To sum up, I would like to propose a few concrete way forward strategies – both for academics and decision-makers who may influence policy and development practice. First, there is an increasing need to prioritize policies to support pastoralists by acknowledging their habitat rights to rangelands. Secondly, indigenous knowledge about their land, its biodiversity – flora and fauna, and their culture is crucial for the pastoralist to maintain their livelihoods. Thirdly, any interventions such as mainstream education should be introduced with a process that integrates the local knowledge and provides learnings that add value to pastoralism without leaving the youths confused and alienated from their identity. The fourth priority is academic, more studies should be done in order to highlight the significant role pastoralists play in maintaining the sustainable management of biodiversity and their continued future role in ecosystem services to counter-argue with the academicians who have a biased myopic view of pastoralism. Lastly, I would like to propose a global exchange of the knowledge of pastoralism and learnings to overcome new 'development' challenges as south–south and north-south multi-dimensional sharing of knowledge.

## ACKNOWLEDGMENTS

I am grateful to all interviewees from Maldhari and Bahima pastoralist tribes, and the entire team of Sahjeevan non-governmental organization, Gujarat, India, and Makerere University, Kampala, Uganda for supporting the author in the data collection phase.

## NOTES

1 The 'Tragedy of the Commons' here refers to the concept from an article by William Forster Lyod (1833) and Garrett Hardin (1968) argued the effect of users who have unregulated grazing common land causing depletion of the resource through their uncoordinated exploitative action.
2 Bhai is referred to in local culture as a brother – a sign of respecting the man. Unfortunately, Salim bhai succumbed to a long illness in 2020.

## REFERENCES

Agrawal, A., and Saberwal, V., 2004. Whither South Asian pastoralism? An introduction. *Nomadic Peoples* 8(3): 36–53.
Awad, M., 1959. Settlement of nomadic and semi-nomadic tribes in the Middle East. *Ekistics* 7(42): 338–343.

Bacon, E.E., 1954. Types of pastoral nomadism in Central and Southwest Asia. *Southwestern Journal of Anthropology* 10: 44–68.

Barnes, A.N., Baasandavga, U., Davaasuren, A., Gonchigoo, B., and Gray, G.C., 2000. Knowledge and practices surrounding zoonotic disease among Mongolian herding households. *Pastoralism* 10: 8.

Clarke, J.I., 1959. Studies of semi-nomadism in North Africa. *Economic Geography* 35: 95–108.

Dyson-Hudson, R., and Dyson-Hudson, N., 1980. Nomadic pastoralism. *Annual Review of Anthropology* 9: 15–61.

Falah, G. 1990. The evolution of semi-nomadism in non-desert environment: The case of Galilee in the 19th century. *GeoJournal* 21(4): 397–410.

Farvar, M.T. 2003. Mobile pastoralism in West Asia: Myths, challenges and a whole set of loaded questions. *Policy Matters* 12: 31–41.

Forhinger, C., Rosqvist, G., Nila, I., and Singh, N., 2021. Reindeer husbandry in peril? How extractive industries exert multiple pressures on an Artic pastoral ecosystem. *People and Nature - A Journal of Relational Thinking* 3: 872–886.

Hardin, Garrett. "The Tragedy of the Commons." *Science* 162 (1968): 1243–1248.

Herskovits, M.J., 1926. The cattle complex in East Africa. *American Anthropologist* 28(1): 230–272.

Little, M. 2016. Pastoralism. In M. Muehlenbein (Ed.), *Basics in Human Evolution*. London: Elsevier, pp. 337–347. https://doi.org/10.1016/B978-0-12-802652-6.00024-4

Muga, G.O., Onyango-Ouma, W., Sang, R., and Afrognon, H., 2021. Indigenous knowledge of Rift Valley Fever among Somali nomadic pastoralists and its implications on public health delivery approaches in Ijara sub-County, North Eastern Kenya. *PLoS Neglected Tropical Diseases* 15:2.

Nori, M., 2022. Assessing the policy frame in pastoral areas of Sub-Saharan Africa (SSA). Policy Paper. RSC PP 2022/03 European University Institute, Italy.

Riggio, J., Jacobson, A.P., Hijmans, R.J., and Caro, T., 2019. How effective are the protected areas of East Africa? *Global Ecology and Conservation* 17: e00573.

Spooner, B., 1971. *The Cultural Ecology of Pastoral Nomads*. Addison-Wesley Module Anthropology 45. Reading, MA: Addison-Wesley, pp. 53.

Sternberg, T., and Chatty, D., 2016. Marginality, climate and resources in pastoral rangelands: Oman and Mongolia. *Rangelands* 38(3): 141–151.

CHAPTER 6

# Factory Schools

## *Erasing Children's Indigenous Knowledge and Languages*

**Geographical Location**: India (Mizoram, Maharashtra), Ecuador (Tena), and Bolivia (Titicaca Lake)

**Indigenous peoples**: Lai tribe, Warli tribe, Huaorani, and Uro indigenous peoples

### 6.1 INTRODUCTION

A news report in the Washington Post (Hedgpeth and Martinez, 2023) indicates that a nonprofit group has identified more boarding schools in the new list released by the National Native American Boarding School Healing Coalition. The news mentions that tens of thousands of American Indian children attended schools that were removed from their land, stripped them of their language, and reprogrammed them under a Manifest Destiny model (ibid). Many tribes in the USA and Canada are investigating what happened to children lost to these schools. Steve Minton (2023) writes that 'The United Nations Declaration on the Rights of Indigenous Peoples delimits indigenous 'sovereignty' to near-tokenism in its insistence on preserving the territorial integrity of nation-states. Instead, what indigenous peoples are being asked to reconcile with is loss, thus cementing the colonization process."

Another news, from India, was about "Outrage from Indigenous leaders: notorious 'Factory School' for 30,000 children set to host World Congress


DOI: 10.1201/9781003153061-6

of Anthropology" to protest against Kalinga Institute of Social Sciences, a residential school for tribal children (Survival International, 2020). What is Factory School? Schrager's article (2018) nicely explains how the education system that we know now evolved only about 200 years ago with the invention of industrialization, which created the need for universal schooling. I quote below from the article (ibid):

> 'Factory Schools' as they are referred to now, originated in early 19th-century Prussia. For the first time, education was provided by the state and learning was regimented. Dozens of students at the time were placed in grades according to their age, and moved through successive grades according to their age, and moved through successive grades as they mastered the curriculum. They took an industrialized approach to education: impersonal, efficient, and standardized.

The idea of a modern-day factory school where a student has to spend their working weekdays in a domestic setting precisely emerges from the industrialized approach to be punctual, docile, and sober. There are many schools of thought, and elite schools, though diverted from the industrialized approach, continue to follow the fundamental rule that students had to be taught in the classroom using a guided curriculum. The question that this chapter aims to explore is how factory school impacts the indigenous knowledge (IK) of forest-dwelling people in the tropics. This chapter will introduce the concept of the language of indigenous communities and explain historically how mainstream schooling erases Indigenous Ecological Knowledge and what it means for the future of Sustainable Development Goals. Factory school is known for their bias in teaching only certain forms and productive modes of knowledge often poses a threat to IK.

Using the comparative case studies from India's Warli tribe and Lai tribe along with Ecuador and Bolivia's indigenous peoples – Huaronis and Uro – respectively, the chapter will explore the risks of factory schooling for the indigenous children and the future of IK. After this brief introduction, the next section will highlight the trend in the non-indigenous education system and the role of language mainstreaming. Thereafter, the empirical qualitative data collected from the three countries, India, Ecuador, and Bolivia, between 2014 and 2018 is presented as qualitative case studies. This is followed by a discussion analyzing the case studies of each indigenous language of four communities. A recommendation for future research agenda is charted out, along with drawing key conclusions.

## 6.2 TREND IN NON-INDIGENOUS EDUCATION SYSTEM AND LANGUAGE MAINSTREAMING

Learning for indigenous peoples is a process. To understand the non-indigenous education system, I propose to use Paulo Freire's reference to a banking paradigm that regards students to be 'adaptable and manageable beings.' According to Freire (1968; see also 1984), the current educational framework is based on the 'Banking Concept of Education' wherein the teacher as a subject is active and the students as passive objects. Freire argues that the 'banking' concept of education is unidirectional; instead of communicating, the teacher 'deposits knowledge' and students only as receivers, filing and storing the deposits. In the banking concept of education, knowledge holders consider the knowledge takers as ignorant thereby making teachers the epistemological authorities in non-indigenous education system. The students' existing knowledge and expertise are not valued putting them at the bottom of the hierarchy of knowledge. Such an education system, when used for mainstreaming children of indigenous peoples, alienates them from their identity and IK by making them adapt to a new language, often in colonial languages, for example, English or Spanish.

The United Nations Declaration on the Rights of Indigenous Peoples (UNDRIP) 2007 recognizes the right to education and languages for indigenous peoples in Articles 13 and 14. The mandate emphasizes that the state should adequately provide access to education for indigenous peoples, including children. One of the highlights is that the education should be provided within indigenous peoples' own culture and should be delivered in their own language. The UNDRIP Articles 13 and 14 are elaborated below:

Article 13

1. Indigenous peoples have the right to revitalize, use, develop, and transmit to future generations their histories, languages, oral traditions, philosophies, writing systems, and literatures, and to designate and retain their own names for communities, places, and persons.

Article 14

1. Indigenous peoples have the right to establish and control their educational systems and institutions providing education in their own languages, in a manner appropriate to their cultural methods of teaching and learning.

2. Indigenous individuals, particularly children, have the right to all levels and forms of education in the State without discrimination.
3. States shall, in conjunction with indigenous peoples, take effective measures, in order for indigenous individuals, particularly children, including those living outside their communities, to have access, when possible, to an education in their own culture and provided in their own language.

The UNDRIP 2007 is a Declaration that was established after decades of indigenous people's advocacy and the United Nations' push for the rights of indigenous peoples. Yet, like many other UN declarations, UNDRIP is non-binding on States. The world's indigenous populations are estimated to be around 350 million or approximately seven and a half billion people who live on this planet and speak about 6,700 languages. However, it is estimated that half of the world's population speaks one or two of the 24 languages, and only 0.1% of the world's population speaks about 3,500 plus of the world's languages (Wilder et al., 2016). About 300 languages, which is half of the world's total languages, are at risk of extinction, and many of these languages are indigenous languages, going extinct one language per 4 months (Rogers and Campbell, 2015).

Indigenous languages and the education system are interlinked. For indigenous peoples, education is reflected in their way of living in different landscapes, cultures, and countries. Indigenous education is oral with storytelling, collaborative learning, and learning by doing based on the wisdom and experience of elders and tradition. One significant difference is that the children are active learners and are encouraged to participate as important members of the teacher-student community. Much of these learnings are indirect and effective for survival, not their own, but also that of the other living beings, including mountains, rivers, and forests that are part of the living being in many indigenous languages. Learning is a process in which the teacher and the student learn together without any hierarchy, unlike the Western learning style. This distinctive feature that Western non-indigenous factory school methods of education disregard traditional indigenous cultures has a negative impact on indigenous students.

The impact of learning a new language for indigenous children in factory school is a big struggle to first learn the foreign language, and then use it to learn new skills. The skills, on the other hand, that the students learn in factory schools are often out of context from their environment and needs. Stripping the basic context and native language of indigenous children has negative implications for their learning of the content about how their indigenous culture is portrayed in the history text and in the language. It also has implications on the way they learn about their own social culture, which is often looked down upon in the modern factory school, particularly lesser-known knowledge of the spiritual understandings and diverse perspectives of governing the forests and its resources with respect and as beings. The biggest drawback of the factory school models mainstreamed for the indigenous peoples across the world using colonial languages to teach is that they hinder them from expressing themselves fully through indigenous languages and cultural practices, alienating them not just from the indigenous language but also from the IK.

## 6.3 CASE STUDIES: LAI AND WARLI TRIBE FROM INDIA, HUAORANIS OF ECUADORIAN AMAZON, AND URO INDIGENOUS PEOPLES OF BOLIVIA

### Data Collection in the Study Areas

From 2014 to 2019, I collected ethnographic primary qualitative data from my fieldwork in the Latin American countries of Bolivia and Ecuador, as well as in India's northeast and western regions. In Bolivia, the data collection spanned over 3 days, but the time to travel from La Paz to Lake Titicaca (Puno) took over 15 days to reach the field site. This was due to roadblocks and protests in La Paz by the labor unions demanding just payment. Despite limited time for actual fieldwork, I gathered all the information during intensive three days living with the Uros (also known as Uru). I was accompanied by a local non-governmental organization (NGO) staff who also acted as an interpreter for languages spoken by Uros (mainly Aymara and Spanish). I stayed in a tourist lodge in the vicinity of the community. The Uros have been exposed to foreigners coming to 'visit and see them,' which made my trip as a normal foreigner curious about their 'way of living.' Given that the NGO staff was also a female author, most of the interviews and community interaction were with female Uro community members. This was in a way ideal for this study because it is the women who took day-to-day care of children's education and language learning.

In Ecuador, the fieldwork was in Tena, Ecuadorian Amazonian region, and I had an opportunity to visit twice; the first trip helped to gain a broad perspective of forest conservation, while the next trip was to hold semi-structured focus group interviews with the Huaorani youths. Due to the remoteness of the location, one of the youths came to pick me from Tena City, and they hitchhiked with wildlife conservation staff. Like Bolivia, the trip to reach the fieldwork site was over seven days since I had to rely on public transportation from Quito. The indigenous youths, along with a local freelance consultant who was also doing forest research in the region, supported me with logistics, including the interpretation. I stayed in Tena City and made the trip a couple of times to meet the community during the fieldwork. During one of the day trips, I shared transportation with the oil corporation explorers, which could have made trouble to establish trust with the Huaoranis, given that they were protesting against any oil pipe passing from the Amazonian forests. However, it was the youth from the community who was accompanying me was the one who organized the trip, which made it easier for me to build trust. Moreover, it was later found out that the youths educated outside acted as mediators for the oil corporations to enter their community.

India's fieldwork in Mizoram's Lai district with the Lai tribe was equally a long journey due to its remoteness like the other two mentioned communities living in isolated landscapes. I audio-video recorded the group discussion and interview to make a documentary film. The biggest challenge was to find someone who could not only understand the almost-lost language of Lai but also translate it into English. The volunteer youth who joined me and the research team from the capital city, Aizawl, was unaware of the language of Lai! Two days were spent finding someone who would be able to help the team interpret the language that was spoken by only a handful of elders. The challenge of this case study was that these handful of men and women speaking Lai were also the ones who resisted speaking or learning Mizo, which meant that finding an interpreter was a difficult task. A retired schoolteacher came to the rescue and acted as a translator. One of the challenges of this case study would be to verify the realistic translation of the narrative. To double-check the data, me and my team in the capital city of Aizwal got a professional translator to help them review the data using audio-visual recordings. The remaining interviews with other Lai people speaking Mizo were easy without any interpretation challenges. The final case study of the Warli tribe from Maharashtra in Western India was also used as a part of the film production therefore all the interviews were recorded in audio-visuals. Both the Lai and Warli tribe's prior informed consent was taken in their own language to record their narratives. The Warli villagers were earlier living in a so-called remote forest mountain landscape, but due to the National Highway that cut through their forests many settlers (non-Warli) have since occupied the area for business or residential purposes. This study area had easier direct access compared to other above-mentioned case study sites.

The comparative case studies are presented with direct quotes using pseudonym names to maintain the anonymity of all interviewees, irrespective of receiving prior informed consent to use their names for the study.

### 6.3.1 Huaoranis of Ecuadorian Amazon, Tena

A relatively small number of households are located in one of the most biodiverse rainforests on the planet. Huaoranis live in the Amazon rainforest of Ecuador near Tena (Map 6.1) and have been displaced earlier due to their protest against oil corporations that drill and install pipelines inside their ancestral land. One of the immediate observations is thatched roofs with minimalist living necessities and basic family farms, a traditional way of living for Huaorani communities. Children live with their parents, learning by using their. Almost everyone knew the skill to navigate a rudimentary canoe – the only way for the people to remain connected using the river with the outside world.

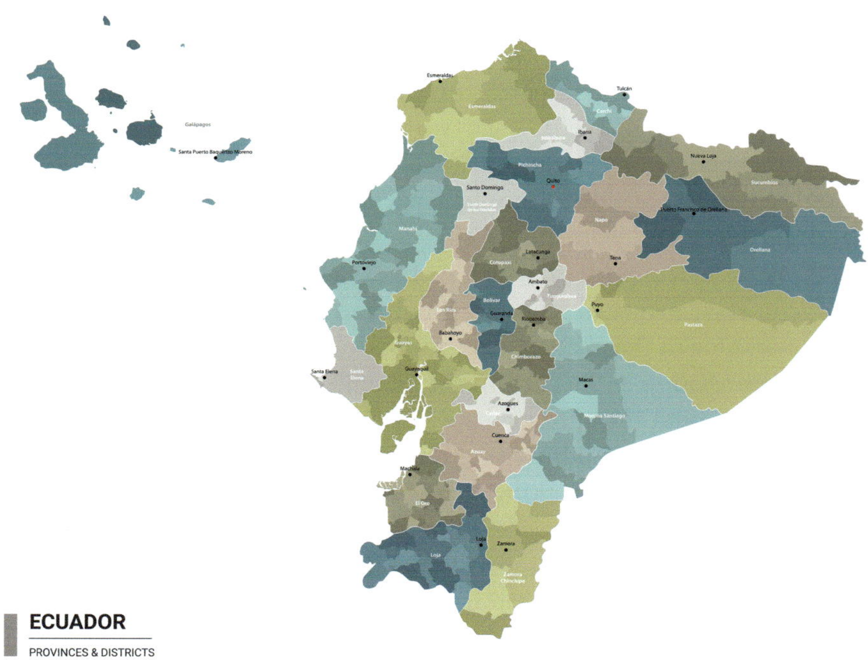


MAP 6.1 Map of Ecuador. Field work area was in the Amazon rainforest in Tena.

Huaoranis came across a factory school education system and a new language when missionaries first approached them. During the focused group discussion, it was evident that the entire community collectively plays an important role in educating the children. One of the elderly women said, "Our language is extinct. Our name is Waodani or Waorani, but in English and Spanish speakers call us Huaorani. For us, rivers, forests, and trees are our life. In our language, we used to have words for forests that mean home. We lost our language therefore the cultural physical and spiritual worlds have now separated putting us at risk of the outside world. With the new language we speak compared to our ancestors, we are losing our detailed knowledge (oral tradition) of ecology, biodiversity, and healing properties. Many children in our community have studied and are studying in cities. What they learn is that money can buy everything."

Some of the community members had planned a hunting trip inside the forest, and I was suggested to join them. The men and women were accompanied by children and youths, which was a traditional way of involving the next generation in learning to hunt. The tradition involved offering to the earth as gratitude for offering bushmeats, the most common wildlife. The education in this way enriches the children with interactive learning and to identify species that are permitted for hunting as per their customary food taboos and other values associated with hunting such as avoiding breeding seasons, or not eating birds of prey. "What we teach our children is that humans are not superior to animals. Hunting tools (preparing the darts using herbal formula) used should not harm or cause pain to the animal. Even the dead animal spirits are alive and therefore a ritual is performed. However, the children who go to school in cities are not taught this respect toward nature. The school education is so different that they (children) feel ashamed of our customs and traditions," says one of the customary leaders who has never been to modern school, but has acquired immense knowledge of trees, medicinal properties, and ecology of the region.

According to the members of the focus group discussion, the youths who have received higher education from the cities have caused harm in many ways including losing interest in their local language over Spanish. The youths have collaborated with the oil corporations and earned a lot of money; while there were also stories about how communities that defied the western schooling have managed to win rights over the forest and against the oil companies. The children in this community did not go to school and were trained by their community members collectively in the local language and using it to learn the names of the trees, plants, herbs, and all botanical and biodiversity knowledge that they use all their senses to identify and recognize to preserve them. The language and traditional knowledge bind them to nature, and they can relate this learning far better than modern school learning, where non-indigenous children learn skills that are not useful to survive in the (Ecuadorian) Amazonian forest.

### 6.3.2 Uro Indigenous Peoples of Bolivia (Andes)

In the Bolivian Andes region, Uro indigenous peoples are known by tourists who visit these remote islands near Lake Titicaca to see the fishing boats made of bundled dried plant totora reeds. For almost 3,000 years, the indigenous peoples, since the time of the Inca civilization, have been living on Lake Titicaca (Map 6.2). The Uru people had almost lost this ancient technique, but it was revived in recent decades, and it is more culture for the tourist and less for the way of living. The Uro people are also known as the people of the water because of their unique culture of living on water and their main source of livelihood has been fishing.

MAP 6.2 Map of Lake Titicaca, the field work area bordering Peru and Bolivia.

I was informed that the Uros ethnic group – the original – might be extinct, and so might the language. The interviewees explained that they might originate from Aymaras and therefore they only speak Aymara and Quechua languages. The study community consisted of only 15 households; however, they were closely knit families, and kinship was strong as a single family. The school is in the highlands, and it is taught in the Spanish language, which has been a cause of resistance by the local communities. "We are ignored by the Bolivian government. Why do we have to learn the non-indigenous way of thinking and living in school and in a language that is colonial?" argued one of the women whose three children are studying in a Christian school, the only source of education on their island.

The focus group discussion with women highlighted the fact that the education they receive from the government does not equip them to continue their indigenous traditional way of living. It fragments the social fabric by ripping the children of their own mother tongue. The children learn Aymara as one of the languages, but the medium of teaching in these schools is in Spanish. The community says the last person to speak some form of the original Uru language is one of the young mothers who argue, "What is the point of education that forces our children to learn a new language and learn about things that matter nothing to our way of living? The weaving and treatment of totora reeds that support our houses, and these floating islands is an art but also a science that children should learn from us. However, our way of living is now only becoming a showcase for foreigners who appreciate it, but that does not help in promoting and sustaining our livelihood that is disappearing, Therefore, the factory school might be helpful to find labour jobs."

The majority of the interviewees explained that even those speaking Aymara are discriminated against despite the fact that the current president (at the time of interviews) belongs to the Aymara ethnic group. None of the interviewees agreed that modern education and a new language are a solution to safeguard their traditional knowledge. On the contrary, the factory school type of education only alienates their knowledge, making it impossible to pass on the oral tradition of community knowledge to the next generation.

### 6.3.3 Lai Tribe of Mizoram, India's Northeast Region

Located in the Northeast region of India, Lawngtalai district of Mizoram, Lai is an autonomous district council (LADC) that is further away from mainland India (Map 6.3). The Lai people are also found in Myanmar.

MAP 6.3 Map of India with highlight of Mizoram state, the study area located in Lawngtalai district.

In India, they share common ancestry with other Northeast Indian tribes. The unique culture of the Lai tribe is the collective kitchen garden, but with education, many families have taken up jobs and invested less time in the forest and agriculture-related activities.

In LADC, almost all schools are run by Christian missionaries teaching in English medium with one subject of the Mizo language. During the field visit, I came across only a few elderly Lai people who continue to speak the language. The majority have adopted the state language of Mizo (another tribe language) along with English. One of the elderly Lai explained, "For the oral traditional stories to sustain the language have to be spoken. You cannot record the IK in another language. It is a piece of knowledge but only half. The remaining half full is lost in the words that often cannot be translated or the expression does not exist in another language. The development has pushed Lai people toward mainstreaming not just adopting the language but with it the culture, the way of living and the narratives that are foreign to us."

During the focus group discussion, many Lai people expressed that the factory school type of education was in earlier days resisted by the local communities. However, resistance meant a lack of support from the government, and mainstreaming was the only option. One of the youths during the discussion highlighted that "Mainstreaming language – everyone speaking only one or two languages – is similar to having a forest with one or two species. In Mizoram, we are rich in biodiverse rainforest, but now mono plantation of rubber trees and oil palm have degraded the entire state of its rich tropical forest. I studied in a residential school, which taught us values very different from the Lai tribe's customs and traditions. Lai tribal children, like me, in residential schools do not get any sense of knowledge about our (Lai) culture or traditional food that is closely linked to forest products. This alienates us from our elders in the community."

### 6.3.4 Warli Tribe of Maharashtra, India's Western Region

My first stop during the fieldwork was at a government-run primary school inside the Warli tribe's village. During the visit, it was evident that the children were rote memorizing the names of the trees, flowers, and birds; a majority of these were not even native. The parents were certainly unhappy and in the focus group discussions expressed "Our children have to deal with two worlds. The first one is real where they live but are distancing due to the school and learning new language (culture). While the second world is that of school – they memorize the names of the birds from the books – none of these birds they have seen or heard, so they have little emotional connection." The community is unimpressed with the school curriculum in the Marathi language because in the history class, the Warli tribe is not mentioned, but rather, all the kings and warriors from the other caste (India has different caste systems) are mentioned. Moreover, the environment that they live in, with high biodiversity of trees and wildlife, is not included in the curriculum.

The observation of the classroom indicated that children were expected to memorize the names of trees and wildlife to qualify for the exam. This, according to the community members, is contrary to their indigenous education, where the emphasis is to identify the trees and wildlife and ability to know how the trees and wildlife grow best (climate, soil, weather, ecosystem, water, and food) and what is the core mutual benefits that they share with the nature. Indigenous education highlights teaching by experiencing it through all its senses and that way, not just memorize the name but also how it is beneficial for health, medicine, food, and fuel for the community. Learning in modern schools lacks these dimensions of core indigenous values about nature and its interdependence.

One of the interviewees summarized, "The school (and teachers) view us as illiterate parents of our children. That is the problem. The view was changed about us when these teachers from urban areas realized they are 'literate' but lacked any sense of connection with nature. We had to teach them how to identify herbs that are natural repellents of insects and bugs and to teach them how to identify forest foods that are highly nutritious. It is only then that the teachers and schools recognized that we carry immense knowledge, which is more valuable than the school certificate. This helped our children to realize that our IK is equally valuable just as the 'modern science.'"

The majority of the school children are from the Warli tribe, which makes it easier for them to identify with each other's culture and 'way of living.' One of the children interviewed by me expressed that she developed interest in environment because of her paternal grandmother, who tells stories and songs to narrate their Warli history; she also learned how their community is the stewards of the forests through the world-famous Warli painting. Children continue with the modern school, but they get additional education about Warli knowledge and are engaged in learning the art of Warli painting – not just how to paint, but also what kind of cow dung mixture with soil becomes the best background of the canvas – that is the walls of their houses.

During the field trip, I had an opportunity to experience a 'Forest Day' where the school organized a mela (festival) for urban dwellers to enjoy local forest food and to mingle with parents and children of Warli who proudly showcase their IK. This annual festival organized by a local Non-Governmental Organization (NGO) aims to 'educate the disconnected urban dwellers' about the roots of Warli culture and its environment. The director of the NGO explained, "When you alienate children from their roots, they will not only begin to integrate the mainstream language and culture, but also begin disliking their indigenous identity. This is politics of education. If you want to displace these communities, bring in Television and Education and new Language for their children. The indigenous children will begin watching tv show and grew up thinking the only best life is in the congested pigeon-hole types of high-rise buildings of the urban metropolitan cities with polluted air. This way the indigenous peoples of future generation are self-displacing and dispossessing."

## 6.4 DISCUSSION

From the above four case studies, the indigenous communities' struggle for ensuring education that is holistic and representative of their culture, languages, and traditional knowledge is evident. The factory type of education system works best for certain class and society, particularly urban industrialized modern society. A blanket approach to implement the mainstream curriculum in the indigenous territories might have a negative impact in the long run for the community, but also for climate adaptation and protection of biodiverse forests.

The four case studies might not be representative of the entire Global South countries, yet it does represent key challenges that many indigenous territories struggle with. The external praise of formal 'factory schooling' as integration in mainstream society ignores the competing indigenous 'way of living' overbearing. As the case studies of Huaoranis, Uros, Lai, and Warli indicate, the decision-making about modern schooling exceeds easy binaries between indigenous and non-indigenous, including the collective indigenous culture vis-à-vis individual rights. For these indigenous peoples, the language and education choices are the results of their decision as to what works best for their family in given insecure forest and land tenure, as well as political conditions.

A call to teach local indigenous languages in school has gained momentum, but that is not easy. In indigenous territories, such as Uros, for a local language to survive revival by introducing it as a language in school will not solve the problem. As indigenous languages are lost, scholars (see Nakata, 2023; Wilder et al., 2016) argue that even IKs are lost along with epistemologies and ontologies that give meaning to all that one knows. Nakata (2023) explains that "These ontologies and epistemologies order our systems of thought, not just what we know but how we think, and how we come to know or believe our knowledge of the world to be valid or true. Without our languages, IKs cannot be expressed in their fullest terms or transmitted to convey their deepest and most intricate meanings."

Evidence from the narratives of the indigenous peoples in the case studies of this chapter is that they are not against education or learning new languages. Rather they argue that education cannot be mainstreamed of what is predominantly catered to urban population and factory school type that has no or little relevance for ensuring their children will become stewards of forests and natural resources. Indigenous language as a language to be taught in school is no solution because language is diverse and holds valuable knowledge associated with modern-day industrial problems. For indigenous peoples, the struggle is natural, like any other population of the world, to learn a new colonial language, but the problem is immense when the language they have known to carry knowledge embedded with it is excluded in the education. The cultural dimension of IK and language could be easily erased with the factory school if there is a lack of appetite for a different 'inclusive indigenous' perspective.

## 6.5 CONCLUSIONS AND RECOMMENDATIONS

The chapter has highlighted that Western education models throughout the world are not a panacea, particularly for acknowledging IK and substituting indigenous education. Further, the case studies of forest-dependent indigenous peoples in Latin America and South Asia showcased that indigenous peoples have many negative experiences that have led to oppression, discrimination, racism, and marginalization of indigenous people. To conclude, it is the lack of environmental justice for indigenous peoples that creates hurdles in understanding that they are guardians of the forests and their immense knowledge. The knowledge of land, governance, language, culture and spiritual ceremony, and kinship cannot be isolated. When they are displaced from their indigenous territories, indigenous men, women, elders, and youths cannot conduct spiritual ceremonies and cultural activities that are the base for their language. From birth to death, ceremonies using forests, water, soil, air, and nature as a whole are the institutions for education for indigenous people. An institution cannot function by taking away culture, ceremony, and language and expecting the indigenous people to continue the knowledge to protect and safeguard their ecosystem and kinship. It will break down and so will the entire ecosystem by opening up space for exploitation of its resources, destruction of its ecosystems, and climate change. As one of the indigenous elders from Ecuador expressed, "We should not be studied, but rather understood. Our language should not be preserved, but rather it should be promoted along with the knowledge and our ways of living. Scientists study the extinction of species across globe and extinction of languages across globe as two different worlds. That's the mistake! The Western science should learn that everything should be understood in totality, and not in an individualistic way."

A few key recommendations for mainstreaming IK and schooling system over the factory school model are (1) for educators developing pedagogy: to learn the local indigenous language before proposing modern school in indigenous territories. There are many schools of thoughts about diversity of pedagogy – from cultural focus pedagogy to localized land-based context-specific pedagogy of indigenous territories; (2) for government decision makers: to learn the wisdom of indigenous peoples – it might not be possible to stop extracting resources, but when West meets as a humble learner of IK there is a window for better middle ground to protect and extract resources; and (3) for scientists, academic scholars and students: to unlearn the discipline expertise and re-learn the art of 'holistic' approach where language and education or forests and climate change or people and development are not an individual entity but part and parcel of same earth and therefore has to be studied as an integrated approach.

## REFERENCES


Freire, P., 1968. *Pedagogy of the Oppressed.* New York: Seabury Press.

Freire, P., 1984. Education, liberation and the church. *Religious Education* 79(4): 524–545. https://doi.org/10.1080/0034408400790405

Hedgpeth and Martinez, 2023. More schools that forced American Indian children to assimilate revealed. *The Washington Post*, accessed via https://www.washingtonpost.com/nation/2023/08/30/indian-boarding-schools/

Minton, S., 2023. We must not forget what happened to the world's Indigenous children. *AEON*, accessed via https://aeon.co/essays/we-must-not-forget-what-happened-to-the-worlds-indigenous-children

Nakata, N.M., 2023. Indigenous languages and education: Do we have the right agenda? *The Australian Education Researcher.* https://doi.org/10.1007/s13384-023-00620-0

Rogers, C., and Campbell, L., 2015. Endangered languages. In M. Aronoff (Ed.), *Oxford Research Encyclopedia of Linguistics.* Oxford: Oxford University Press. Accessed via https://oxfordre.com/linguistics/display/10.1093/acrefore/9780199384655.001.0001/acrefore-9780199384655-e-21;jsessionid=EC8F13ACF077CA86646304308C3D4735

Schrager, A., 2018. The modern education system was designed to teach future factor workers to be "punctual, docile, and sober", accessed via https://qz.com/1314814/universal-education-was-first-promoted-by-industrialists-who-wanted-docile-factory-workers

Wilder, B., O'Meara, C., Monti, L., and Nabhan, G., 2016. The importance of Indigenous knowledge in curbing the loss of language and biodiversity. *BioScience* 66(6): 500.

CHAPTER 7

# Indigenous Peoples and Wildlife Coexistence in Tropical Mountains

## *Socio-Cultural Impact*

**Geographical Location:** India (Changthang, Tibetan Plateau in Ladakh), Nepal (Annapurna Conservation Reserve) and Colombia (Sierra Nevada de Santa Marta)

**Indigenous Peoples**: Changpa indigenous people, Gurung and Magars of Annapurna Range, and Kogi and Arhuaco of the Sierra Nevada de Santa Marta

## 7.1 INTRODUCTION

Globally, mountain regions are fragile ecosystems. Indigenous peoples and local communities living in and around protected forest and land areas including in mountain regions are exposed to greater levels of day-to-day interactions with wildlife. Due to protected area designation, the conservation practice is often introduced by the Western science model, which often calls any aggravating interactions between humans and wildlife a 'human-wildlife conflict.' There is a risk in how the Western science narrative has used the term 'human-wildlife conflict' as a blanket approach to promoting wildlife conservation through the lens that all humans are


DOI: 10.1201/9781003153061-7

threats to wildlife. Often natural science scholars through quantitative data project numbers of human-wildlife conflicts and their impact, for example, Choudhury's (2004) study highlights between 1980 and 2003 more than 1,150 humans and 370 elephants died due to the conflicts in India's Northeast states. The Western narrative using numbers and statistics has simply ignored the traditional coexistence of indigenous peoples and wildlife. The Western conservation narrative continues to use a 'blanket' definition of 'humans' in the human-wildlife conflict debate, undermining the role of local communities and indigenous peoples but also ignoring the root cause of any such conflicts. The studies use conflicts to highlight how such conflicts are indicators that new forms of enforcement of environmental protection are needed for the conservation of wildlife. Jorgensen et al.'s (1978) study on Black Bears in the 1970s blames humans – without specifying if these were local communities or poachers – for killing the wildlife and thereby transforming the wildlife habitats. Such discourses were the reason for increased policies and regulations that introduced the laws of making it illegal to kill animals, as well as making it socially unacceptable.

In recent decades, the human and wildlife narratives continued to fragment human and wildlife, stating their 'competitive' relationship, and as Hill (2004) points out, human-wildlife conflict turned into a political conflict between farmers/people and between institutions. A growing number of studies highlight that the human-wildlife is a reciprocal process wherein humans and animals are negatively affected by the conflict. Many studies point out that the major cause of human-wildlife conflict is increasing human populations adjacent to wildlife habitats. According to König et al. (2020), the literature on human-wildlife conflicts has grown in the last two decades, mainly looking at the negative outcomes for people and wildlife. Yet, these studies have failed to capture how the communities have been managing the wildlife co-adapting to live in shared landscapes. The focus has remained on reducing negative interactions rather than promoting the examples of positive coexistence and cohabitation that exist around the world. The reason for such a low number of positive human-wildlife coexistence empirical studies could be linked to the way wildlife conservation theories have been constructed and studied by scholars promoting fortress conservation – keeping humans away from the wildlife, and wildlife is protected in a reserved nature without any human interactions. Yet increasingly it is clear that 'pristine wilderness,' a natural zone free of humans, as a conservation technique is a failure.

Fletcher et al. (2021), in their study entitled, indigenous knowledge and the Shackles of Wilderness, argue that a pristine wilderness where rich biodiverse forests continue to exist without humans present is a Eurocentric construct. They argue that it is an idea that gained significance during European colonial and conservation efforts in the 19th and 20th centuries in Asia, Africa, and the Americas, and it may be visible in the agendas of many international conservation organizations, governments, and some philanthropists. In a policy brief on the Post-2020 Global Biodiversity Framework, the United Nation's special rapporteur on human rights and environment, David Boyd, explains that 'fortress conservation' is motivated by the mistaken belief that successful conservation outcomes require 'pristine wilderness' free from human inhabitants. Pristine wilderness, from the lens of indigenous peoples, would mean ecologically degraded and culturally poor without the presence of humans. Degradation is a result of unsustainable management of natural resources combined with climate change including drought, flood, and desertification. Ecosystem degradation reduces biodiversity and impacts wildlife. Therefore, wilderness discourse in conservation practices has only helped further to disempower indigenous peoples and local communities who are the stewards of ecosystems and landscape management.

This chapter examines three different cases – given that there are no single indigenous or traditional knowledge systems – that provide empirical qualitative evidence on the ways indigenous peoples collect and transmit knowledge that is critical for maintaining the well-being of people as well as all the living beings of the landscapes.

### 7.1.1 Definitions

In this chapter, the term human-wildlife conflict will not be used because the way it is defined is problematic. Yet, the chapter does not deny the existence of conflicts but argues that the reasons and causes for an increase in conflict need to be understood and acknowledged and that there is a lack of management of resources. The World Wildlife Fund report (Gross et al., 2021) explains that human-wildlife conflict is "when encounters between humans and wildlife lead to negative results, such as loss of property, livelihoods, and even life." The report (ibid) refers to "'human-wildlife conflict' as to struggles that arise when the presence or behavior of wildlife poses actual or perceived direct, recurring threats to human interest or needs, often leading to disagreements between groups of people and negative impacts on people and/or wildlife."

This chapter will examine the other side of the coin of human-wildlife conflict and that is the human-wildlife coexistence. The definition by Gross et al. (2021) of "human-wildlife coexistence refers to people and wildlife existing in proximity to each other, whether in contentious, neutral, or beneficial coexistence." The underlying aim is that "the interests and needs of both humans and wildlife are met and is characterized by a level of tolerance on the human side" (ibid). The term 'tolerance,' the report explains, is the "passive or active acceptance of a wildlife population, and it depends on the risk-benefit beliefs people have towards a species." This chapter will also argue that the social science lens should be used to understand the human-wildlife relationship and that the term 'human-wildlife conflict' should be replaced by the term 'human-wildlife coexistence.'

## 7.2 THE STUDY AREA AND RESEARCH METHODOLOGY

The primary data collection was done by me between 2013 and 2018 in three countries – Nepal, Colombia, and India. I did solo trekking of the mountains in Nepal and India, while in Colombia I was accompanied by a colleague. The study areas are comparative yet distinct in characteristics of indigenous knowledge, and the indigenous peoples in each of the study areas have maintained the landscape and wildlife along with local livelihoods, customs, culture, and languages.

I spent approximately 20 days in each of the study areas. The temperature in the study area ranged from –30°C for the Changthang desert region just like Artic cold to –10°C in the Annapurna region while pleasant 10 degrees Celsius in the Sierra Nevada de Santa Marta. The average altitude was 5,500 m (18,000 ft) in Changthang, while in Annapurna, I did a study of 4,000 m, and in the Sierra Nevada de Santa Marta, the altitude was 5,000 m. In all the study areas, I had research assistants from the local area who helped as guides and language transcribers. I stayed with the indigenous peoples and traveled for two days with nomadic Changpa women herding goats and sheep on mountains.

The data were collected through primary methods. Primary qualitative data was collected from the indigenous peoples using semi-structured interviews and field observation. The interviews were also conducted in the form of Focus Group Discussions (FGDs) in all study areas to examine the human-wildlife coexistence, the local philosophy for coexistence, management, and risk with mountain livelihood. Key informant interviews were conducted with selected informants while in FGDs, I was the key facilitator, and respondents had dialogues among themselves as well as with me. It provided an opportunity to interact with everyone in the community and gain information about indigenous knowledge and their perception of human-wildlife coexistence. I did transect walks (walk in the mountains for observation) to observe and reconfirm the responses from the semi-structured interviews. To prepare for the field trip to collect primary data, I did secondary data collection from sources such as books, the internet, reports, and peer-reviewed articles. The data was analyzed using a simple descriptive qualitative method.

## 7.3 CASE STUDIES: THE NATURE OF HUMAN-WILDLIFE COEXISTENCE

### 7.3.1 Ladakh's Changtang Tibetan Plateau, India

The Tibetan plateau of eastern Ladakh, a Northern state of India, is known as Changtang (Map 7.1). The economy of the Changpa is based on livestock. The nomadic pastoralists of Changtang move from one season to another, covering the mountain range during winter and valley during the short summer period. Migratory routes are already established, and they follow them every year – often, traveling together as group members of hamlets. The summer encampments have stone walls, and the shelter uses tents made of Yak wool, while the winter camps have a room with a stove and roof.

The cover photo of this book is of the Changpa people who are posed together with me in the Changthang region during my fieldwork trip.

Traditionally, due to nomadic livelihood, Changpa men and women are extremely gender balanced and independent. During the winter season, the women take their goats and sheep to higher altitudes in search of fodder – they walk alone on different routes and spend 5–6 hours before returning to their base. One of the women explained, "We have about 200 goats, known as Changthangi, and they graze in the cold, arid region of Changthang at an altitude of 3000 meters. For us, the trade of these animals is important, particularly the unprocessed cashmere traded that fetches us enough money to buy basic goods. Yet, when a snow leopard takes away one of our animals, we do not get angry because like us they have the right to eat. Our elders taught us that we are sharing the land of snow leopards, wolves, and other big animals and birds and they are the

MAP 7.1 Map of Ladakh.

first ones to enjoy the benefits of land before us (humans). I learned that the snow leopards also respect us because they maintain their boundary and do not attack humans. This mutual sense of respect existed for hundreds of years, and we would like to continue this way."

One of the elder men, leader of the hamlet, questioned why the people (researchers and practitioners) who have never lived together with wildlife have the power to decide and conclude that indigenous peoples like Changpas have conflicts with animals. He said, "When I was young, the snow leopards were visible regularly and we youngsters would play while they would wander around, without troubling us. Just like the mountain goats and wolves and other predator animals would do. However, with tourists from France and other places who pay a lot of money to 'see' the snow leopards things have changed for us. Some of the outsiders have convinced a few of the youngsters Changpas to release a goat to attract snow leopards. This is not our practice. By feeding the goats, these outsiders are changing the lifestyle of the leopards and thereby impacting our livelihoods. Because these youngsters chase away other animals who in turn become aggressive – these animals try to attack us sometimes thinking we will hurt them. Therefore, we no longer allow any tourists to visit our area to watch the snow leopards."

During the group discussion, many have voiced that all animals that are roaming freely are treated with respect. In Buddhist practice, these wild animals are considered part of the ecosystem, and no one hunts or hurt them. The Changpas do not claim compensation from the government for the loss of their animal because, as per indigenous customs, that is the payment or rights of the predators who in turn guard their landscapes. The Chagpas expressed that they value the food chain – whereas, in snow, leopards are an integral part of it, along with other animals – therefore, they must maintain the food chain.

One of the fears among Changpas is that soon, with the demarcation of their pastureland in the Tibetan plateau, they may be forced to become settled pastoralists, and they may lose the guardianship of protecting and conserving their landscapes including the wildlife. Losing the right to wildlife and ecological conservation, they fear to eventually lose the rich indigenous knowledge.

### 7.3.2 Sierra Nevada de Santa Marta, Colombia

The Kogi and Arhuaco indigenous peoples are special because they are one of the few who have maintained a way of living by harming no one (Map 7.2). The Kogis and Arhuacos were interviewed by me during the year 2013–2014 in the Sierra Nevada of Santa Marta, which is in Northern Colombia. It is a special range because it is the highest coastal range in the tropics and is highly biodiverse, with almost 40 endemic species of Colombia found here including other wildlife such as tapir, jaguar, cougar, deer, and Santa Marta parakeet. The national park is a tourist attraction. After all, it offers a diverse range of climate, flora and fauna, and terrain, and it is a unique location in the world because it is not connected to the Andes. Approximately 30,000 indigenous peoples live here, and the Tairona culture is promoted through UNESCO's Biosphere Reserved. The International Union for Conservation of Nature (IUCN) also regarded it as one of the world's irreplaceable protected parks that hosts some of the threatened species.

One of the Kogi youths, using his poporos, explained their relationship with the coexistence of the wildlife as part of their blood, which has continued since the pre-Colombian era. He explained, "We strongly believe in the Great Mother as the creator figure of nature while we are her children

MAP 7.2 Map of Santa Marta, Colombia.

who are protected from being exploited and destroyed. Therefore, the community strongly follows Mamos, who are highly respected Kogis and believe that harmony in the world is the key aspect. To preserve our traditional way of life, we do not interact with the modern world, and we weave our clothes and eat food that is grown here. We never hunt or kill wildlife, but with the drug dealers encroaching on our land, we are facing conflicts with them, but not with the wild animals. For use, it is human-human conflict – good vs bad – one who protects the wildlife vs one who kills for making money."

During the focus group discussion, Arhuacos explained that they do not value material goods, and therefore, they only sell a limited version of their hand-woven bags woven by women to buy basic household food. The indigenous peoples are often killed by drug dealers who aim to expand their territory of drug dealing and plantations of marijuana in the wildlife-protected areas. One of the Arhuacos women explained, "Traditionally we have two houses one for cooler weather and the other warmer – that way we are indirectly interacting with a diverse range of wildlife over a year. Often, government officials blame us or make us the scapegoat for the killing of wildlife by an outsider. They even threaten us to burn down the coca plantations which we do not produce for commercial reasons."

The indigenous peoples in this region regard themselves as guardians of the forests and wildlife biodiversity, and therefore, they often protest against projects such as ecotourism and hydroelectric because they bring more destruction to their protected areas. Therefore, they prefer to remain isolated, and the protected area would enable them to protect and coexist with the wildlife.

### 7.3.3 Annapurna Conservation Reserve, Nepal

The primary economy of the Annapurna Conservation Reserve is ecotourism (Map 7.3). The Gurung and Magars indigenous people are dependent on tourism for livelihoods. Degradation of the ecosystem without any restoration projects, the human-wildlife conflict perception is that the indigenous peoples are responsible for the decline. The Gurung people consider the Annapurna range and all wildlife inside it as sacred and have remained strict with the outsiders or tourists as well. Though they are the victim of the Conservation Area project because they no longer can collect firewood or do domestic grazing, they do, however, still protect the wildlife.

During the focus group discussion, it became evident that the majority of the indigenous people living in this mountain range find the Western model of wildlife conservation problematic. One of the elder indigenous

MAP 7.3 Map of Annapurna range, Nepal.

people mentioned, "When we have traditionally been living in this land how come the Western and outsiders (urban-based Nepalis) have more knowledge than us to decide how our land should be protected? This is insane because our indigenous knowledge is ignored completely. We are not consulted on how to coexist with the wildlife rather we are excluded from the decision-making process of conservation. This external system will not work, and it will destroy our all knowledge because if we do not practice, we cannot teach our children. Our knowledge and wildlife will disappear together."

In the Annapurna range, the indigenous peoples are facing the threat of eviction from the conservation area and do not gain enough economic benefit from wildlife-related tourism. A woman during the interview replied, "How do the policymakers function is beyond my comprehension. They are the outsiders they should be thrown out, but we who have protected the forests and wildlife are threatened. Our conflict is with the outsiders just like the wildlife we are treated badly. As a child, I used to see so much wildlife, but not with settler communities in the region we are seeing an increasing number of human-wildlife conflicts because the settler communities hold no indigenous knowledge to protect and conserve the wildlife."

## 7.4 DISCUSSION: LISTENING TO INDIGENOUS VOICES INDIGENOUS PEOPLES AND THEIR WILDLIFE COEXISTENCE

From the above findings, it is clear that the perception of indigenous peoples about their wildlife coexistence is that the danger is from other external humans. The three case study areas are considered to be 'pristine' according to indigenous peoples, but they all fear that their ancestral territories that have been long shaped and managed using indigenous knowledge predate all fortress conservation enclosures by several millennia. One of the common factors of conflicts is not wildlife, according to the respondents, it is about the lack of acknowledging the indigenous knowledge by the international institutions who attempt to integrate them into mainstream conservation plans.

The possible solution therefore could be that (1) human-wildlife coexistence territories should be legally recognized as autonomous indigenous people and local community-led and managed territories; (2) need for active voice and transparent engagement of indigenous peoples' knowledge in adapting the Western model in the tropical forest countries; (3) recognize and secure future tenure rights to land, sub-soil resources, diverse livelihood and support indigenous rights to protect and coexists with wildlife; and (4) expansion with settler communities in and around protected areas would become zones of conflict and retaliatory killing of wildlife, including highly valued carnivores such as snow leopards and jaguars by tourism, therefore, need to identify drivers of such conflicts rather than punishing the indigenous peoples and locals of depriving their rights to livelihoods.

The Kogi community in Colombia has been promoting bridging the knowledge gap by developing teaching materials – including conservation techniques – with Western scholars and trainers so that the Kogi methods will be beneficial for other wildlife territories. The argument put forth by the Changpas is that in indigenous territories, about 60% of all species on Earth exist, which is a good indicator that the only way for wildlife to thrive is through indigenous knowledge. The criticism of Western science and the Western idea of conservation stems from the fact that it has not been tested and practiced for generations or millennia, unlike indigenous knowledge that exists because it has withstood all sorts of challenges.

Listening is an integral part of the indigenous knowledge. The Kogis, like other indigenous peoples of this study, are being trained to be a good listener. Amplifying their voice is secondary to them because, as a good listener, one can make good decisions based on what they heard from Mother Earth and all its living beings. For all the indigenous peoples interviewed in this study from Nepal, Colombia, and India, the wildlife conservation process is not about rewilding of the landscaped or restoration or regeneration, which are the popular conservation methods, but instead, there is a need to let the wildlife have the same respect and space as a human being. Indigenous knowledge must be remembered over time and be made known to a wider population; that way it continues to flourish.

## 7.5 CONCLUSIONS WITH RESEARCH AND POLICY RECOMMENDATIONS

The three study areas of this chapter are located in diverse ecologies, and landscapes have similarities in the way indigenous knowledge has promoted human-wildlife coexistence.

### 7.5.1 Recommendations for Key Multi-stakeholder Groups

#### *7.5.1.1 Research Institutes and Researchers*

The key recommendation for research institutes and researchers is to promote indigenous scholars in empirical research and writing. There is a lot written about human-wildlife interactions, but an overwhelming majority of these publications are by Western non-indigenous scholars. This needs to be changed to ensure equitable perspective and participation in research by indigenous scholars. Next, to move away from the traditional negative 'natural science' quantitative data narrative of 'human-wildlife conflict' toward adopting a better understanding of the positive discourse of 'human-wildlife coexistence.' In doing so, the linkage between the integration of human-wildlife coexistence into the implementation of the Sustainable Development Goals framework is needed to ensure wildlife conservation from human dimensions. The other key recommendation is to work with social science scholars to add qualitative data and transdisciplinary research to add social and cultural aspects of indigenous people's successful traditional coexistence and practice of wildlife conservation.

#### *7.5.1.2 Policymakers*

For better implementation of living in harmony with nature, there is a need for decision-makers to adopt human-wildlife coexistence as a target of the Convention on Biological Diversity process. To acknowledge the indigenous knowledge in wildlife conservation and coexistence and involve indigenous peoples as key actors.

#### *7.5.1.3 Private Companies and Enterprises*

To avoid promoting and advertising wildlife tourism that picks up one wildlife, such as snow leopards, thereby ignoring the significance of the other ecosystems and wildlife. This selective process is problematic because the tourists ignore or devalue the role of other wildlife in nature. For a long-term sustainable solution, private companies need to recognize the human rights of marginalized groups living in vulnerable regions (Bose, 2017) as critical for wildlife conservation.

#### *7.5.1.4 Philanthropists*

To commission research to identify successful stories and processes of human-wildlife coexistence in indigenous territories and promote awareness campaigns against negative narratives for indigenous people's wildlife conservation efforts. To ensure the funding directly reaches the target audience such as that of indigenous peoples.

### 7.5.2 Indigenous Peoples and Local Communities

To highlight the traditional knowledge that encourages promoting a holistic understanding of wildlife conservation. The world needs to learn that indigenous peoples in many parts of the world, as seen in the case studies of this chapter, live harmoniously with a diverse range of wildlife and have well-established oral traditions and cultural practices, which help them learn to coexist. The crux of human and wildlife coexistence is not 'development' but rather leading a minimalist way of living that promotes the rights of all living beings.

### 7.5.3 Media and Communication Organization

To cautiously highlight the 'human-wildlife conflicts' in social media such as films and print media that provide a holistic analysis of why the conflict happened and what are the underlying causes. The media's unbiased reporting such as why wildlife, for example, elephants damage the crops of farms, would educate the general public in better understanding of the connections of human-wildlife conflicts linked with the degradation of protected areas due to the construction of roads or loss of food due to invasive species, or poverty, inequality, lack of livelihood and land tenure rights.

### 7.5.4 Civil Societies

To provide capacity building for a diverse range of stakeholders that promote the concept of coexistence over conflicts and thereby acknowledge that indigenous knowledge like Western science can provide organizational support. Another recommendation would be to ensure gender and diversity equity and respect human rights in human-wildlife coexistence.

### 7.5.5 Government Authorities

To promote transparency by adopting free, prior, and informed consent and an inclusive, participatory engagement of the communities traditionally living in and around wildlife in diverse ecosystem landscapes. To address the threats of coexistence from unsustainable development, food insecurity, climate crisis, and fortress conservation that hinders or ignores indigenous knowledge.

## REFERENCES

Bose, P., 2017. Climate adaptation: Marginal populations in the vulnerable regions. *Climate and Development* 9(6): 575–578.

Choudhury, A., 2004. Human-elephant conflicts in Northeast India. *Human Dimensions of Wildlife* 9: 261–270.

Fletcher, M.S., Hamiltan, R., Dressler, W., and Palmer, L., 2021. Indigenous knowledge and the shackles of wilderness. *Proceedings of the National Academy of Sciences* 118(40): e2022218118.

Gross, E., Jayasinghe, N., Brooks, A., Polet, G., Wadhwa, R., and Hilderink-Koopmans, F., 2021. *A Future for All: The Need for Human-Wildlife Coexistence*. Gland, Switzerland: WWF.

Hill, C.M., 2004. Farmers' perspectives of conflict at the wildlife-agriculture boundary: Some lessons learned from African subsistence farmers. *Human Dimensions of Wildlife* 9: 279–286.

Jorgensen, C.J., Conley, R.H., Hamilton, R.J., and Sanders, O.T., 1978. Management of black bear depredation problems. *Proceedings of the Eastern Workshop on Black Bear Management and Research* 4: 297–321.

König, H.J., Kiffner, C., Kramer-Schadt, S., Fürst, C., Keuling, O., and Ford, A.T., 2020. Human-wildlife coexistence in a changing world. Special section: Challenges of and solutions to human-wildlife conflicts in agricultural landscapes. *Conservation Biology* 34(4): 786–794.

CHAPTER 8

# Communicating the Art of Bridging Indigenous Knowledge with Science and Policy

**Geographical Location:** India, Nepal, Indonesia, Uganda, Colombia, Peru, Mexico, and Kenya

## 8.1 INTRODUCTION

As seen in the previous chapters of this book, the emerging global 'development' pressure and historical dispossession are faced by indigenous communities from extractive industries for transition to a 'green,' low-carbon economy. Transition minerals such as lithium, nickel, cobalt, copper, etc., for development are found in traditional indigenous territories that have social and environmental impacts on indigenous peoples. Tropical deforestation is rapidly increasing every year. According to recent data from the World Resource Institute's Global Forest Watch platform, the tropics lost almost four million hectares (or about ten million acres) of forest in 2022, which is higher from 3.75 million hectares (or about 9.3 million acres) in 2021. Scientists across the globe argue that forest loss in tropical countries has led to the release of over two gigatons of carbon dioxide into the atmosphere and has led to either an increase



DOI: 10.1201/9781003153061-8

or decrease in precipitation that may be linked to drought, flood, and fire. Further, the loss of forests in the tropics has affected the wildlife population. Estrada et al. (2022) study findings indicate that about 30% of indigenous people's lands are primate range, and they safeguard 71% of the world's primate biodiversity therefore safeguarding indigenous people's lands, languages, and cultures represents the chance to prevent the extinction of the world's primates. The recent COVID pandemic is also linked to increased trading of wildlife and bushmeat, which could spread viruses and cause pandemic outbreaks. In previous chapters, irresponsible mining showed it can contaminate rivers and fish. In other words, forests in the tropics have a chain reaction. Whatever happens there in the forests, does not remain limited to local impact there but impacts the entire world. One of the chapters also examined how indigenous knowledge has been compared with scientific knowledge as traditional vis-à-vis modern science, which creates a problem for indigenous children and youths who, through 'modern' education systems, are separated from their environment. The lack of indigenous education has affected the cultural sustainability and future of communicating indigenous knowledge with the next generations.

### 8.1.1 What is Communication Science?

For many, particularly non-social science background, science communication is all about how you talk or communicate your message to others. However, this notion is currently changing and is challenged by many. Communicating science used to be complicated and often considered a specialized subject by a few of those who are experts in this field. It is broadly understood as a range of activities that connects science with 'people.' The term 'people' refers to a diverse range of audiences – young to elders and urban to rural and indigenous territories. Therefore, the crucial goal of communicating science or science communication is to simplify scientific findings to inform non-experts with the primary intention of raising awareness, building a collective understanding of scientific findings, and engaging in a dialogue to find solutions for societal challenges (Bose, 2019). In one way, the term 'science communication' could indicate that it is a one-directional flow of information – from the haves to the haves not – often assuming that the audience receiving the information are not experts, but they could also be equal partners such as expert to expert dialogue. On the other hand, the communicators of science are often the 'literates' of Western education and use different communication modes such as storytelling, films, cartoons, posters, or even more recently social media to capture their audience's attention. With growing criticism about Western science and science communication many practitioners and scholars are shifting the trend toward being inclusive by listening to non-scientists and no longer shaming them as ignorant (see Orthia, 2020).

Many communities in the Global South have historically colonized and been subjected to injustices for centuries of resource extraction in Africa, Asia, and Latin America. There are similarities in the way Western science is conducted by excluding marginalized groups and the strategy used for science communication that is top-down linked to colonialism. Both create inequities in the production of scientific research – by excluding communities of indigenous peoples, people of color, black, and Afro-Colombians, LGBTQ+ communities, or even people who are differently abled – to the outputs or the way it is communicated in the English language. The English language used in most scientific communication has limited the audience and further exploitation of using indigenous knowledge without giving credit to intellectual knowledge.

Modern-day science communication calls for acknowledging the resilience, resistance, and demand for dignity that is crucial for bridging Western and indigenous science by promoting horizontal knowledge of co-creation. The perspective of the Global South as the primary focus rather than a Global North approach to communicating science demands transparency and seriousness. The new generation of communicating science demands an initiative that is entirely from the South, without any predesigned Western protocols or frameworks of doing science or the findings that are adapted to be politically correct to avoid touching topics of justice, decolonization, or racial discrimination. What this new generation of communicating science does differently is interact and bridge South-South collaboration from diverse ethnic groups, countries, and cultures. Collectively dedicate time and resources to creating visual aids by considering the needs of multi-stakeholder perspectives and communicating in multilingual modes of communication. The new generation of science communication works with transparency that indigenous knowledge is localized therefore 'factory production of communication' to scale up the solutions for context-specific problems will not work.

## 8.2 COMMUNICATING INDIGENOUS KNOWLEDGE: EUROCENTRISM AND DECOLONIZATION

In 2019, at the International Union of Forest Research Organization World Congress in Curitiba, Brazil, my keynote speech highlighted the recent trends in science communication and its potential and threats to indigenous knowledge (Bose, 2019). Bose (2019) points out that in the Global South, scientists and science communicators are acknowledging the importance of indigenous knowledge for socially inclusive narratives. Yet, the challenge they face for transformative efforts – moving from the science communications that are Eurocentric toward indigenous knowledge – is that indigenous knowledge is not treated as scientifically proven or relevant knowledge. Persistence underestimation of indigenous knowledge systems misses out on the otherwise value they would have added to science communication uptake, even for the Global North. Much data and evidence that have emerged from research over the last five decades clearly indicated that there is a need for an enabling environment, inclusive participation, and political will to achieve the United Nations Sustainable Development Goals. Nevertheless, the practice on the ground continues to be the opposite, where the science communication is designed in a way to exclude those who do not speak English, Spanish, or French – the colonial languages – further alienating those who did not go to school or had formal modern education as 'illiterate.'

The latest trend in indigenous studies has shown that a majority of scholars writing about 'decolonization' are those from the Western world further reinforcing the Western science perspective of what decolonization theory should be and how it should be done – not to mention that it rarely highlights the need to determine indigenous peoples and those who have been colonized should be at the center stage in decision-making of such research. This Eurocentric research agenda combined with the top-down funding to do research on decolonization by Western scientists is both symptomatic of indigenous people's alienation as well as a detrimental of creating social distance in communicating knowledge-building inequalities.

One of the most classic examples of international development going wrong is communicating climate science. For many people, climate science is all about what they read or hear about on the Intergovernmental Panel on Climate Change (IPCC). The IPCC is regarded as having a great influence over many governments' decision-making on climate change, particularly in terms of funding. Yet, the IPCC clearly had missed out on including indigenous knowledge for a long time by only focusing on Western scientific research. At the time of writing this chapter, there has been no inclusion of indigenous authors in any of the IPCC chapters. This is similar to the findings that a majority of authors writing on decolonization are Western scholars from colonist countries and without any collaboration with authors from countries that were colonized. Appropriation of knowledge in Western science derails the way indigenous peoples' contribution to thinking and understanding gets ignored in IPCC reports.

The scholars working for Western institutions often lack the bigger picture of what it means to be excluded due to their colonial privileges. Many Western institutions are recruiting scholars from the Global South to overcome the gap of narrative that Westernized science often misses such as interacting with indigenous communities to ask what knowledge they have about the social, cultural, ethical, and biophysical knowledge of their environment. It is expected that these interactions will aid scientists in Western countries in learning the indigenous languages and knowledge that has been documented at the local level. These observations and rich data in the form of oral tradition are valuable to predict the future normal or abnormalities about weather, farming and fishing production, breeding season, foraging, and hunting activities, path to consider for nomadic pastoralism, and any pandemic outbreaks. Communication about this indigenous knowledge and values has been critical to addressing environmental and social problems.

By recognizing decolonization efforts, which in principle is a genuinely significant effort, but unfortunately fails to touch upon the core issue of political-economic challenges. The reason for the failure of the otherwise well-intended 'decolonization' theory is that it has primarily focused on the historical wrongdoings of the past such as the injustices of colonial empires. The colonial empire, though real and responsible for the economic mismatch with the Global North, is in a powerful position compared to the Global South. This economic disparity makes it even tougher when the discussion about 'decolonization' is initiated and monopolized by the Global North countries. The Eurocentric narratives produced by Western scholars about decolonization are one-sided – lack the collaboration of indigenous scholars – and often, postcolonial countries with rich environmental resources are forced to accept the funding for conservation that further alienates the indigenous peoples from their land and resources. Indigenous knowledge, therefore, when treated within the limits of any Western science frameworks, would create miscommunication.

The Western view of the environment is communicated from a Eurocentric lens as a resource and commodity to be exploited while the indigenous view of the world is communicated as a common where humans, wildlife, and nature are part of a single body – a web of life – constantly communicating to maintain the balance.

## 8.3 HOW SCIENCE COMMUNICATION INFLUENCING POLICY?

This section shares the evidence from qualitative data gathered during the fieldwork to examine how policymakers use knowledge to make policy decisions, what kind of knowledge they have, where they access that knowledge, and from whom and why. Drawing experiences using semi-structured interviews with forest and land-use policymakers from eight countries of the Global South – India, Nepal, Indonesia, Uganda, Colombia, Peru, Mexico, and Kenya – are used as the qualitative data to analyze how and what kind of knowledge is being used to influence the environmental and social decision.

### 8.3.1 Methodology

For empirical evidence, I collected the qualitative primary data from eight countries in the Global South. The interviewees were either audio or audio-visual recorded during the interview with prior informed consent. All of the interviews were conducted in English, except for two interviewees who were in Spanish and Bahasa Indonesia. At the time of interviews between 2015 and 2018, the interviewees were elected political leaders at the state, or national level with decision-making authority about the environment and forests within their jurisdiction in indigenous territories. In total, 24 interviews were conducted in the eight countries mentioned above. For ease of presentation, interviews are presented as a regional case, but they do not represent the opinion of the entire country or the continent. The below presentation highlights the semi-structured interview questions such as how policymakers use knowledge to make policy decisions, what kind of knowledge they have, and from where they access that knowledge, from whom, and why.

### 8.3.2 Asia: India and Nepal

Eight elected politicians, of whom seven men and one woman, were interviewed in India and Nepal. All of them have been in politics and in powerful positions related to forests, water, and the environment. The woman politician from India had been involved in historical forest rights policymaking that gave recognition to the forest-dependent Adivasis or indigenous peoples. In Nepal, one of the politicians is active in promoting wildlife conservation and considers indigenous peoples' participation as enriching for preserving and promoting indigenous knowledge and their ways of governing resources. None of the policymakers interviewed had ever read academic papers, and the reasons were varied. For some lack of access to scientific journals was the challenge, but for everyone, the common reason was that scientific papers are meant for students as a textbook and have no use for politicians.

All respondents explained that local knowledge is the key source, and it includes talking to farmers and indigenous communities. According to one of the elected representatives, science communication undermines the existing knowledge and treats the receivers – the policymakers – as amateurs. He explained, "Western scholars come here (India) and collect data without even knowing any Indian languages. Then, they reproduce the data that we have provided without involving any co-authors from our country. The final conclusion and analysis lack a cumulative understanding of culture, historical, and political scenarios – rather it is more for their scientific career that it benefits. How much scientific knowledge when communicated in an alien language will benefit us (our policy)?"

### 8.3.3 East Africa: Kenya and Uganda

Seven elected politicians were interviewed, and all of them have been active in the field of natural resources, pastoralism, and landscape restoration. The information they gather is from their constitutions – rural and indigenous territories – and that helps them set the agenda. Although two of the leaders have studied abroad and were familiar with scientific journals, none of them valued peer-reviewed publications. The reason is that the information scientific literature provides is either basic or lacks sustainable recommendations. Almost all of them agreed that communicating science should be a bottom-up approach, but in practice, it is a mismatch with the politics of power to promote development. Science communication therefore is valuable but plays a secondary role in making any impact. The knowledge of weather, wildlife, and way of living by the pastoralist groups in Kenya and Uganda, through their protest and voices, ensures that they are being heard. Although there are many instances where indigenous knowledge is cosmetically appreciated, it rarely gets acknowledged in policymaking.

### 8.3.4 Latin America: Colombia, Peru, and Mexico

Eight of the politicians interviewed four were women in leadership positions and some of them were elected for second or third term. One of the women politicians from Colombia explains, "Our knowledge from Amazon to Andes region has contributed largely to Western science to produce modern-day medicine. Why do we need to rely on Western science communication?" Science communication in Latin America is not beneficial due to many political respondents finding that the biggest challenge is the English language. One of the challenges, as mentioned by a respondent in Mexico, is that politicians are bombarded with information, which runs the risk of propaganda and/or industry-led investment agenda. The respondents often rely on media communication of science for policy decision-making due to the simplified version of complicated jargon that science produces.

From the above empirical evidence, one can safely conclude that among politicians and policy decision makers science communication is not the primary source of information.

## 8.4 ART OF BRIDGING WESTERN SCIENCE AND INDIGENOUS KNOWLEDGE

In summary, a few key take-home messages from the findings of this chapter's key research questions are the following:

Communicating science is productive when delivered in local languages. The policymakers need scientific communication which is the bottom-up approach.

Policymakers do not refer to scientific research and science communication because it does not cater to their needs, and it fails to provide a cumulate understanding of the problem.

One of the risks for science communication is that the western transformed version of the narrative gets the privilege of being the only (valid) narrative produced by science communicators.

Alternatively, the solution lies with the local and indigenous communities in the Global South. Decolonizing research entails that indigenous scholars and wisdom holders lead the (Western) science themselves, incorporating diverse cultures into science communication. Science communicators need to expand their horizons and include cross-cultural relations of different types of knowledge – indigenous and Western science to open conservation of bridging gaps. Doing so will benefit not only the local context but also multi-stakeholder and would impact international policy and practice paradigms, improving indigenous rights, strengthening governance, and acknowledging indigenous knowledge.

## REFERENCES

Bose, P., 2019. Communicating with 'people' in scientific forestry discourse. Keynote speech at the International Union of Forest Research Organization, Curitiba, Brazil. https://www.purabibose.com/communication/communicating-with-people-in-scientific-forestry-discourse

Estrada, A., Garber, P.A., et al., 2022. Global importance of Indigenous Peoples, their lands, and knowledge systems for saving the world's primates from extinction. *Science Advance* 8(32): Article eabn2927.

Orthia, L.A., 2020. Strategies for including communication of non-Western and indigenous knowledges in science communication histories. *Journal of Science Communication* 19(02): A02. https://doi.org/10.22323/2.19020202

# Index

adaptation 8, 20, 40, 81
adaptive 1, 23, 65
agriculture 8, 40, 63, 78
  farm (includes farmers and farming) 7, 14, 31, 33, 39, 41–49, 54, 59–61, 63–64, 74, 85, 96, 102, 104

belief 1, 2, 10, 14, 15, 23, 39, 86–87
Bolivia 11, 38, 41–42, 47–49, 68–69, 72–73, 76–77

case studies 3–4, 10, 24–26, 35, 42, 48, 50, 55–56, 69, 72, 74, 80–82, 88, 96
climate 6, 15–16, 20, 48–49, 58–59, 79, 81–83, 86, 90, 101
coal 24, 32, 34–35, 68
coexistence 84–97
collective 1, 3, 5, 6, 15–18, 23, 25–26, 30, 33–35, 41, 43, 46–51, 56–57, 59, 61, 63, 75, 78, 81, 99, 100
Colombia 8, 11, 14, 18, 22–25, 29–32, 35–37
communication 1, 7–11, 96, 99, 100–106
conflict 10, 25, 31–32, 54, 61, 84–89, 92–96
conservation 2, 4, 14, 16, 19, 27–28, 34, 39, 41, 45–46, 48–49, 56–60, 63, 72–73, 84–86, 90–97, 104, 106
cultural 1, 4, 6, 14–16, 19, 23, 38, 39, 48–51, 53, 58, 70, 72, 75, 81, 82, 84, 86, 95, 96, 99, 102, 105

dances 1, 16
decolonization 10, 100, 101
deforestation 15, 18, 23, 26, 28, 32, 36, 44, 98
dryland 1, 11, 52

ecological knowledge 1, 2, 14, 48–49, 69
ecology 19, 61, 75
economic 2, 4, 10, 16, 19, 22, 23, 26, 35, 39–41, 53–54, 56, 93, 102
Ecuador 11, 47, 68–69, 72, 74, 76, 82
environment 1–2, 5, 8, 11, 15, 18–20, 23, 26, 34, 38, 39, 46, 49, 50, 54, 57, 72, 79, 80, 82, 85, 86, 96, 99, 101–103
environmental defenders 5, 18–20, 23, 26
ethics 24, 59
ethnographic 3, 4, 24, 35, 45, 55, 72

forest 1, 3, 5–8, 10, 15, 17–21, 23, 24, 26–28, 31–35, 37, 41, 44–46, 49, 50–51, 54–59, 69, 71–76, 78–84, 86, 92–94, 98, 99, 101, 103, 104
FPIC (Free, Prior and Informed Consent) 4, 34–36, 51

gender 7, 47, 88, 97
Global North 20, 100–102
Global South 8, 65, 81, 100–103, 105
gold 24, 26–31, 35–37
governance 20, 23, 40, 48, 50, 59, 61, 63, 82, 106

history 1, 8, 29, 59, 61, 63, 72, 79, 80
hunting-gathering 26, 28, 44–46, 49, 57, 75, 102

impact 5, 23, 26, 36, 41, 44, 48, 54, 58, 60, 64, 69, 71, 81, 84–86, 90, 98, 99, 104, 106
India 4, 7–8, 11, 14, 19, 22–25, 32–33, 35–36, 38, 41–42, 48–53, 55–57, 59–60, 65, 68–69, 72–73, 77, 79, 84–85, 87–88, 95, 98, 103, 104

indigenous communities 1, 3, 5, 10, 11, 14–18, 26, 28, 29, 32, 33, 38, 39, 41, 48, 50, 51, 69, 80, 98, 102, 104, 105
indigenous knowledge (IK) 3–21, 23, 27–36, 39–50, 53–61, 64–65, 69–70, 72, 86–106
Indonesia 11, 14, 38, 41, 44, 49, 98, 103

Kenya 7, 53–54, 103–104
knowledge (including Indigenous knowledge, IK) 1–35, 39–51, 53–54, 56–105

land 1, 3, 5–11, 17–19, 23, 26–35, 46–51, 54, 55, 60, 63, 65–66, 68, 74, 81, 82, 84, 88, 89, 92, 93, 96, 99, 102
  landscapes 5–11, 18, 20, 23, 28, 41, 48, 52, 59, 62, 64, 71, 73, 85–87, 90, 95, 97
  land-use 19, 23, 31, 45, 51, 54–56, 61, 63, 103, 104
landscapes 10, 11, 18, 23, 41, 48, 52, 59, 63, 64, 71, 73, 85, 86, 90, 95, 97
language 1–3, 5–8, 10, 11, 24, 26, 28, 42, 47, 56, 61, 68–83, 87, 99–105

Mexico 98, 103, 105
mining 4, 19, 22–37, 44, 48, 61, 65, 85, 99
mountain 1, 3, 6, 34, 42–43, 47, 53, 71, 73, 84, 85, 87–89, 91–93, 95, 97
music 1, 16, 56

natural resources 3, 19, 22, 29, 81, 86, 104
Nepal 11, 84, 87, 93, 95, 98, 103, 104

oral tradition 1, 14, 17, 39, 58, 70, 75, 77, 78, 96, 102

paintings 1, 80
pastoralist 3–4, 6–7, 52–67, 88, 90, 104
Peru 11, 22, 24–27, 35, 47, 98, 103, 105
policy 11, 19–20, 24, 36, 39, 44, 51, 54–56, 66, 86, 93, 95, 98, 103
poor 5, 27, 54, 61, 86
  poverty 8, 50, 96
practice 1–2, 7, 10, 14–17, 19–20, 22–23, 26, 31, 34–35, 39–40, 42, 51–54, 56, 59, 61, 63, 65–67, 72, 84, 86, 89–90, 93–96, 101, 104, 106

qualitative methods 3, 10, 88

rivers 1–3, 6, 8, 17, 23, 24, 26–28, 40, 41, 43, 71, 74, 75, 94, 99
rural 3, 29, 99, 104

science 2, 3, 5–8, 11, 13–17, 19–20, 39, 40, 51, 64, 69, 77, 80, 82, 84, 85, 87, 94, 95, 97, 98–106
social 2–9, 15, 16, 23, 26, 39, 50–51, 53, 61, 69, 72, 77, 85, 87, 95, 96, 98–103
songs 1, 16, 56, 58, 80
stakeholders 6, 7, 25, 28, 29, 35, 60, 63, 97
story 34
  and storytelling 58, 71, 100
Sustainable Development Goals 20, 69, 95, 101

tenure (includes forest tenure and land tenure) 5, 11, 19, 23, 26, 44, 49, 51, 59, 60, 81, 94, 96
tradition (includes traditional and traditionally) 1–3, 6–8, 10, 13–19, 21, 23, 26–39, 41–45, 47–54, 56–59, 63–66, 70–71, 74–80, 85, 86, 88, 91–93, 95–99, 102
tribal 7, 19, 69, 79
  Adivasis 7, 11, 32, 104
  tribe 3, 11, 14, 19, 21–22, 32, 38, 52, 56, 61, 63–64, 68–69, 72–73, 77–80

Uganda 11, 14, 38, 41–42, 45–46, 49, 52, 55–56, 60–63, 66, 98, 103–104
urban 1, 18, 41, 58, 63, 79–81, 93, 99

Western 1–9, 11, 13, 15–17, 19–20, 39, 45, 51, 55–59, 61–62, 64–65, 71–73, 75, 79, 82, 84–85, 93, 95, 97, 100–106
wildlife 24, 32, 34–35, 68
women 5, 19, 25, 27, 29, 31–33, 35–36, 42–43, 45–48, 55–56, 61, 64, 72–73, 75, 77, 82, 87–88, 92, 105

For Product Safety Concerns and Information please contact our EU representative GPSR@taylorandfrancis.com Taylor & Francis Verlag GmbH, Kaufingerstraße 24, 80331 München, Germany